A HISTORY OF LIFE ON EARTH

A HISTORY OF LIFE ON EARTH

UNDERSTANDING OUR PLANET'S PAST

The Changing Earth Series

JON ERICKSON

☑®

Facts On File, Inc.

AN INFOBASE HOLDINGS COMPANY

A HISTORY OF LIFE ON EARTH: UNDERSTANDING OUR PLANET'S PAST

Facts On File, Inc.
460 Park Avenue South
New York NY 10016

Library of Congress Cataloging-in-Publication Data

Erickson, Jon, 1948–
     A history of life on earth : understanding our planet's past /
Jon Erickson.
          p. cm. — (The changing earth series)
      Includes bibliographical references and index.
      ISBN 0-8160-3131-2 (acid-free paper)
      1. Historical geology. I. Title. II. Series: Erickson, Jon.
Changing earth.
QE28.3.E75 1995
560—dc20   94-25290

Facts On File books are available at special discounts when purchased in bulk quantities for businesses, associations, institutions or sales promotions. Please call our Special Sales Department in New York at 212/683-2244 or 800/322-8755.

Text design by Ron Monteleone/Robert Yaffe
Jacket design by Catherine Rincon Hyman
Printed in the United States of America

RRD FOF 10 9 8 7 6 5 4 3 2 1

This book is printed on acid-free paper.

CONTENTS

TABLES

ACKNOWLEDGMENTS

The author thanks the following organizations for providing photographs for this book: the Chicago Field Museum of Natural History, the National Aeronautics and Space Administration (NASA), the National Optical Astronomy Observatories (NOAO), the National Museums of Canada, the National Park Service, the U.S. Geological Survey (USGS), and the Woods Hole Oceanographic Institution (WHOI).

INTRODUCTION

The most fascinating field of geology is the study of our planet's past. The history of the Earth is written in its rocks and the history of life on Earth is told by its fossils. Earth history is divided into units of geologic time according to the type and abundance of fossils in the strata. The fossil record provides valuable insights into the evolution of the Earth. Knowledge of the origination and extinction of species throughout geologic time is also necessary for understanding the evolution of life.

When fossils are arranged chronologically, they vary in a systematic way according to their positions in the geologic column. This observation led to one of the most important and basic principles of historical geology, whereby periods of geologic time can be identified by their distinctive fossil content. Geologists are thus able to recognize geologic time periods based on groups of organisms that were especially abundant and characteristic during a particular time. The occurrence of certain organisms defines each period, and the succession of species is the same on every continent and is never out of order.

These principles became the basis for establishing the geologic time scale and ushered in the birth of modern geology. Nineteenth-century geologists in Great Britain and Western Europe delineated the major periods by using fossils to define the boundaries of the geologic time scale. The entire geologic record was created using relative dating techniques, which placed geologic time units in their proper order without reference to their actual age. Later, as radiometric dating techniques became available, geologists applied absolute dates to units of geologic time, further perfecting their understanding of Earth history.

The development of a geologic time scale applicable for the entire world requires that rocks from one locality be correlated with those of similar age elsewhere. A comprehensive view of the geologic history of a region is obtained by correlating rocks from one location to another over a wide area.

Geologic maps display the distribution of rock formations and present the composition, structure, and geologic age of rocks, factors that are essential for constructing an accurate history of the Earth.

The geologic time scale is divided into eons, which are the largest divisions and up to a billion or more years in length. They include the Archean and Proterozoic eons, also called the Precambrian and known as the age of primitive life, and the Phanerozoic eon, known as the age of advanced life. The Phanerozoic is further broken down into eras, including the Paleozoic or the age of ancient life, the Mesozoic or the age of middle life, and the Cenozoic, known as the age of new life.

The eras are divided into periods, which take their names from the localities with the best rock exposures. Each period is characterized by less profound changes in organisms than the eras, which mark the boundaries of extinctions, proliferations, or rapid transformations of species. The two periods of the Cenozoic are subdivided into epochs due to the greater historical detail provided by more recent rocks. The epoch we live in is called the Holocene, which is synonymous with the beginning of civilization.

1

PLANET EARTH

O ur sun, with its nine planets and their satellites, is a rarity among stars, one of only a small number of single, medium-sized stars in our galaxy. The inner rocky, or terrestrial, planets are much alike, with the notable exception of the Earth, the only planet with a water ocean and an oxygen atmosphere. It is also the only terrestrial planet with a rather large moon, a pairing that still defies explanation.

The atmosphere and ocean evolved during a tumultuous period of crustal formation, volcanic outgassing, and comet degassing. Numerous giant meteorites slammed into the Earth, adding unique ingredients to the boiling cauldron. Raging storms brought deluge after deluge and unimaginable electrical displays. Out of this chaos came life.

THE SOLAR SYSTEM

Some 15 billion years ago, the universe originated with a force whose power is still hurling the farthest galaxies away from us at nearly the speed of light. The steady expansion of the beginning universe might have been temporarily sped up by a sudden inflationary bulge, as the new universe rapidly

Figure 1-1 The Andromeda Galaxy is one of the nearest spiral galaxies to our Milky Way. Courtesy of NOAO

ballooned outward for an instant and then settled down to a steady growth. As the protouniverse expanded, it cooled sufficiently to allow basic units of matter to clump together to form billions of galaxies each containing billions of stars (Fig. 1-1). The first galaxies evolved when the universe was about 1 billion years old and only about a tenth of its present size.

Our Milky Way galaxy has five spiral arms that peel off a central bulge. New stars originate in dense regions of interstellar gas and dust called giant molecular clouds. Several times a century, a giant star over 100 times larger than the sun explodes, producing a supernova a billion times brighter than

TABLE 1–1   SOME CHEMICAL ELEMENTS*

Name	Symbol	Atomic Number	Atomic Weight
Hydrogen	H	1	1
Helium	He	2	4
Oxygen	O	8	16
Carbon	C	6	12
Neon	Ne	10	20
Nitrogen	N	7	14
Magnesium	Mg	12	24
Silicon	Si	14	28
Iron	Fe	26	56
Sulfur	S	16	32
Argon	Ar	18	40
Aluminum	Al	13	27
Calcium	Ca	20	40
Sodium	Na	11	23
Chromium	Cr	24	52
Phosphorus	P	15	31
Manganese	Mn	25	55
Chlorine	Cl	17	36
Potassium	K	19	39
Titanium	Ti	22	48
Cobalt	Co	27	59
Nickel	Ni	28	59
Zinc	Zn	30	65
Fluorine	F	9	19
Copper	Cu	29	64
Vanadium	V	23	51
Scandium	Sc	21	45

*Elements are listed in descending order of their relative abundance in the universe. That is, in this table Hydrogen is the most abundant element, Scandium the least abundant.

an ordinary star. When a star reaches the supernova stage, after a very hot existence spanning several hundred million years, the previously stable nuclear reactions in its core become explosive. The star sheds its outer covering, while the core compresses to an extremely dense, hot body called a neutron star; this would be like squeezing the Earth down to about the size of a golf ball.

The expanding stellar matter from the supernova forms a nebula composed mostly of hydrogen and helium along with particulate matter that comprises all the other known elements (Table 1-1). About a million years later, the solar nebula collapses into a star. Shock waves from nearby supernovae compress portions of the nebula, with gravitational forces causing the nebular matter to collapse upon itself, forming a protostar. As the solar nebula collapses, it rotates faster and faster, and spiral arms peel off the rapidly spinning nebula to form a protoplanetary disk. Meanwhile, the compressional heat initiates a thermonuclear reaction in the core, and a star is born.

A new star forms in the Milky Way galaxy every few years or so. About 4.6 billion years ago, our sun, an ordinary main-sequence star, ignited in one of the dusty spiral arms of the galaxy. Single, medium-sized stars like the sun are a rarity, and due to their unique evolution such stars appear to be the only ones with planets. Thus, of the myriad stars overhead, only a handful might possess a system of orbiting planets, and fewer yet might contain life.

During the sun's early developmental stages, it was ringed by a protoplanetary disk composed of several bands of coarse particles, called planetesimals (Fig. 1-2), accreting from grains of dust cast off by a supernova. Some 100 trillion planetesimals orbited the sun during the Solar System's early stages of development. As they continued to collide and

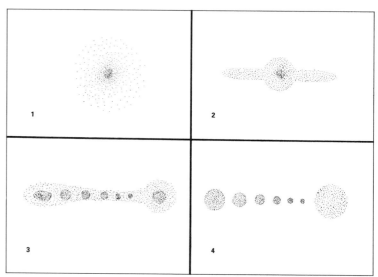

Figure 1-2 The formation of the Solar System from planetesimals in the solar nebula.

grow, the small rocky chunks swung around the infant sun in highly elliptical orbits along the same plane, called the ecliptic.

The constant collisions among planetesimals built larger bodies, some of which grew to over 50 miles wide, but most of the planetary mass still resided in the small planetesimals. The presence of large amounts of gas in the solar nebula slowed the planetesimals, enabling them to coalesce into planets. The planetesimals in orbit between Mars and Jupiter were unable to combine into a planet due to Jupiter's strong gravitational attraction and instead formed a belt of asteroids, many of which were several hundred miles wide.

The Solar System itself is quite large, consisting of nine known planets and their moons (Fig. 1-3). The image of the original solar disk can be traced by observing the motions of the planets, all of which revolve around the sun in the same direction it rotates, and all but one, Pluto, within 3 degrees of the ecliptic. Some 7 billion miles from the sun lies the heliopause, which marks the boundary between the sun's domain and interstellar space. About 20 billion miles from the sun is a region of gas and dust, possibly remnants of the original solar nebula. Astronomers think a belt of comets lying on the ecliptic exists in this region. Several trillion miles from the sun is a shell of comets that formed from the leftover gas and ice of the original solar nebula.

Figure 1-3 **A photo montage of the Solar System, showing Earthrise over the lunar surface, along with Mercury, Venus, Mars, Jupiter, and Saturn.** Courtesy of NASA

THE PROTOEARTH

As the Earth continued to grow by accumulating planetesimals, most of which had temperatures exceeding 1,000 degrees Celsius, its orbit began to decay due to drag forces created by leftover gases in interplanetary space. The formative planet slowly spiraled inward toward the sun, sweeping up additional planetesimals along the way like a cosmic vacuum cleaner. Eventually, the Earth's path around the sun was swept clean of interplanetary material, and its orbit stabilized near its present position.

The core and mantle segregated possibly within the first 100 million years, during a time when the Earth was in a molten state heated by radioactive isotopes and impact friction from planetesimals. The presence of magnetic rocks 2.7 billion years old suggests the Earth had a molten outer core comparable to its present size at an early age. The Earth's interior was hotter, less viscous, and more vigorous, with a highly active convective flow. Heavy turbulence in the mantle with a heat flow three times greater than today produced violent agitation on the Earth's surface. This turmoil created a sea of molten and semimolten rock broken up by giant fissures, from which fountains of lava spewed skyward.

The early Earth did not possess an atmosphere to hold in the internally generated heat, and the surface rapidly cooled, forming a thin basaltic crust similar to that of Venus. Indeed, the moon and the inner planets offer clues to the Earth's early history. Among the features common to the terrestrial planets was their ability to produce voluminous amounts of basaltic lavas. The Earth's original crust has long since disappeared, remixed into the interior by the impact of giant meteorites that were leftovers from the creation of the Solar System.

The formative Earth was subjected to massive volcanism and meteorite bombardment that repeatedly destroyed the crust. A massive meteorite shower, consisting of thousands of 50-mile-wide impactors, bombarded the Earth and its moon between 4.2 and 3.8 billion years ago. The other inner planets and the moons of the outer planets show dense pockmarks from this invasion (Fig. 1-4). The meteorite bombardment melted large portions of the Earth's crust, nearly half of which contained large impact basins up to 10 miles deep.

As the meteorites plunged into the planet's thin basaltic crust, they gouged out huge quantities of partially solidified and molten rock. The scars in the crust quickly healed, as batches of fresh magma bled through giant fissures and poured onto the surface, creating a magma ocean. The continued destruction of the crust by heavy volcanic and meteoritic activity explains why the first half-billion years of Earth history are missing from the geologic record.

Figure 1-4   Multiple impact craters on the ancient surface of Saturn's moon Rhea. Courtesy of NASA

THE MOON

A popular explanation for the creation of the moon supposes a collision between the Earth and a large celestial body. According to this theory, soon after the Earth's formation, an asteroid about the size of Mars was knocked out of the asteroid belt either by Jupiter's strong gravitational attraction or by a collision with a wayward comet. On its way toward the inner Solar System, the asteroid glanced off the Earth (Fig. 1-5), and the tangential collision, which lasted for half an hour, created a powerful explosion equivalent to the detonation of an amount of dynamite equal to the mass of the asteroid. The collision tore a huge gash in the Earth, and a large portion of its molten interior along with much of the rocky mantle of the impactor spewed into orbit, forming a ring of debris around the planet called a protolunar disk.

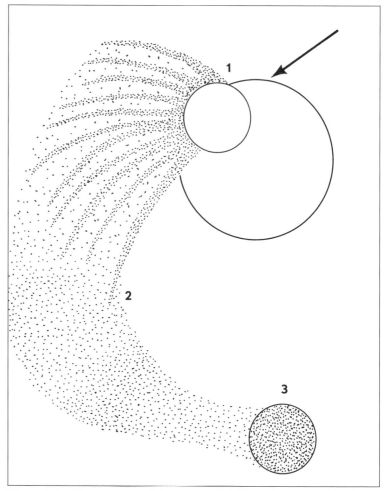

Figure 1-5   The creation of the moon: 1) A Mars-size asteroid collides with the Earth; 2) material splashes up into orbit around the planet; and 3) it coalesces into the moon.

The force of the impact might have knocked the Earth over, tilting its rotational axis about 25 degrees. Similar collisions involving the other planets, especially Uranus, which orbits on its side like a rolling bowling ball, might explain their various degrees of tilt and elliptical orbits. The glancing blow also might have increased the Earth's angular momentum (rotational energy) and melted the planet throughout, forming a red-hot orb in orbit around the sun. The present angular momentum of the Earth suggests that other methods of lunar formation such as fission, capture, or assembly in place were unlikely.

The new satellite continued growing as it swept up debris in orbit around the Earth. In addition, huge rock fragments orbited the moon and crashed onto its surface. The massive meteorite shower that bombarded the Earth equally pounded the moon, and numerous large asteroids struck the lunar surface and broke through the thin crust. Great floods of dark basaltic lava spilled onto the surface, giving the moon a landscape of giant craters and flat lava plains called maria from the Latin word for "seas" (Fig. 1-6).

The moon became gravitationally locked onto its mother planet, rotating at the same rate as its orbital period, causing one side to always face the Earth. The moon exerts a force on the spinning Earth called nutation

Figure 1-6 The lunar terrain, showing numerous meteorite craters and expansive lava plains. Courtesy of NOAO

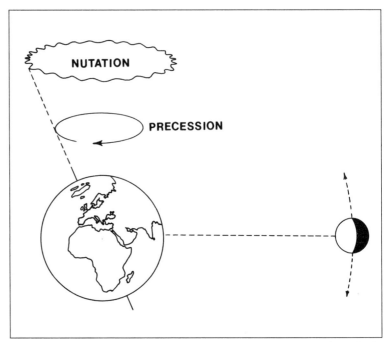

Figure 1-7 The Earth's precession (wobble of the Earth's axis) is caused by the gravitational pull of the moon, whose variable motions produce nutation (nodding of the Earth's axis).

(Fig. 1-7) that causes the rotational axis to precess, or wobble. Many moons around other planets share similar characteristics with the Earth's moon, suggesting they formed in the same manner. Since the Earth's sister planet Venus evolved under similar circumstances and is so much like our planet, the absence of a Venusian moon is quite curious. It might have crashed into its mother planet or escaped into orbit around the sun. Perhaps Mercury, which is about the same size as the Earth's moon, was once a moon of Venus.

The early moon orbited so close to the Earth it filled much of the sky. The presence of a rather large moon, the biggest in our Solar System in relation to its mother planet, the two forming a twin planetary system, might have had a major influence on the initiation of life. The unique properties of the Earth-moon system raised tides in the ocean, and cycles of wetting and drying in tidal pools might have helped the Earth acquire life much earlier than previously thought possible.

THE ATMOSPHERE

For the first half-billion years, while the Earth was spinning wildly on its axis and surface rocks were scorching hot, the planet lacked an atmosphere and was enveloped in a near vacuum much like the moon is today. Soon after the meteorite bombardment began about 4.2 billion years ago, the Earth acquired a primordial atmosphere composed of carbon dioxide, nitrogen, water vapor, and other gases spewed out of a profusion of volcanoes. The atmosphere was so saturated with water vapor, atmospheric pressure was nearly a hundred times greater than it is today. Some meteorites that hit the Earth were stony, composed of rock and metal, others were icy, composed of frozen gases and water ice, and many contained carbon (imagine millions of tons of coal raining down from the heavens). Perhaps these carbon-rich meteorites bore the seeds of life, which might have existed in the universe eons before the Earth came into being. Comets, composed of ice and rock debris, also plunged into the Earth, releasing large quantities of water vapor and gas. These cosmic gases were mostly carbon dioxide, ammonia, and methane.

Most of the water vapor and gases originated within the Earth. Magma contains large quantities of volatiles, volcanic gases comprising mostly water and carbon dioxide, which make it more fluid. Tremendous pressures deep inside the Earth keep the volatiles within the magma, and when the magma rises to the surface, the drop in pressure releases the trapped water and gases, often explosively. The early volcanoes erupted violently because the Earth's interior was much hotter than it is now, and the magma contained higher amounts of volatiles.

Oxygen originated directly by volcanic outgassing and meteorite degassing and indirectly by the breakdown of water vapor and carbon dioxide by

Figure 1-8 An artist's rendition of the Venus rift valley, which at 3 miles deep, 175 miles wide, and 900 miles long, is the largest canyon in the Solar System. Courtesy of NASA

the sun's strong ultraviolet radiation. All oxygen generated in this manner quickly bonded to metals in the crust, much like the rusting of iron. Oxygen also recombined with hydrogen and carbon monoxide to reconstitute water vapor and carbon dioxide. A small amount of oxygen might have existed in the upper atmosphere, where it provided a thin ozone screen. This would have reduced the breakdown of water molecules by ultraviolet rays and prevented the loss of the Earth's water, a fate that might have visited Venus eons ago (Fig. 1-8).

Nitrogen, which comprises about 80 percent of the present atmosphere, originated from volcanic eruptions and the breakdown of ammonia, a molecule with one nitrogen atom and three hydrogen atoms, and a major constituent of the primordial atmosphere. Unlike most other gases which have been replaced or recycled, the Earth retains much of its original nitrogen. This is because nitrogen readily transforms into nitrate, which easily dissolves in the ocean, where denitrifying bacteria return the nitrate-nitrogen to its gaseous state. Decaying organisms also release nitrogen back into the atmosphere. Therefore, without life, the Earth would long ago have lost its nitrogen and possess only a fraction of its present atmospheric pressure.

THE OCEAN

While the atmosphere formed, the Earth's surface was constantly in chaos. Winds blew with a tornadic force, and fierce dust storms on the dry surface blanketed the entire planet with suspended sediment much like the Martian dust storms of today (Fig. 1-9). Huge lightning bolts flashed across the sky, and

Figure 1-9   The surface of Mars, showing boulders surrounded by wind-blown dust and sand. Courtesy of NASA

the thunder was earth-shattering as one gigantic shock wave after another reverberated over the land. Volcanoes erupted in one giant outburst after another. The sky lit up from the pyrotechnics created by the white-hot sparks of ash and the glow of constantly flowing red-hot lava. The restless Earth was rent apart as massive quakes cracked open the thin crust. Huge batches of magma flowed through the fissures and flooded the surface with voluminous amounts of lava, forming flat featureless plains.

The intense volcanism lofted millions of tons of volcanic debris into the atmosphere, where it remained suspended for long periods. Ash and dust particles scattered sunlight and gave the sky an eerie red glow like that on

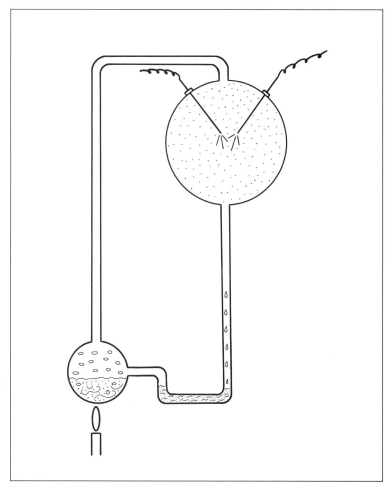

Figure 1-10 Spark discharge chambers simulate prebiotic conditions on the early Earth by recreating an ancient ocean, atmosphere, and lightning storms.

Mars. The dust also cooled the Earth and acted as particulate matter, upon which water vapor could coalesce. When temperatures in the upper atmosphere lowered, water vapor condensed into clouds. The clouds were so thick and heavy they almost completely blocked out the sun, and the surface was in near darkness, dropping temperatures even further.

As the atmosphere continued to cool, huge raindrops fell from the sky, and the Earth received deluge after deluge. Raging floods cascaded down steep mountain slopes and the sides of large meteorite craters and gouged out deep canyons in the rocky plain. When the rains ceased and the skies finally cleared, the Earth emerged as a giant blue orb, covered by a global ocean nearly 2 miles deep and dotted with numerous volcanic islands.

Ancient marine sediments found in the metamorphosed rocks of the Isua Formation in southwestern Greenland support this scenario for the creation of the ocean. The rocks originated in volcanic island arcs and therefore lend credence to the idea that plate tectonics operated early in the history of the Earth. They are among the oldest rocks, dating to about 3.8 billion years ago, and indicate that the planet had surface water by this time.

TABLE 1–2   THE EVOLUTION OF THE BIOSPHERE

Event	Billions of Years Ago	Percent Oxygen	Biologic Effects	Results
Full oxygen conditions	0.4	100	Fishes, land plants, and animals	Approach present biologic environs
Appearance of shelly animals	0.6	10	Cambrian fauna	Burrowing Habitats
Metazoans appear	0.7	7	Ediacaran fauna	First Metazoan fossils and tracks
Eukaryotic cells appear	1.4	>1	Larger cells with a nucleus	Redbeds, multicellular organisms
Blue-green algae	2.0	1	Algal filaments	Oxygen metabolism
Algal precursors	2.8	<1	Stromatolite mounds	Initial photo-synthesis
Origin of life	4.0	0	Light carbon	Evolution of the biosphere

During the years between the end of the great meteorite bombardment and the formation of the first sedimentary rocks, vast quantities of water flooded the Earth's surface. Seawater probably began salty, due to the abundance of chlorine and sodium provided by volcanoes, but did not reach its present concentration of salts until about 500 million years ago. The warm ocean was heated from above by the sun and from below by active volcanoes on the ocean floor, which continually supplied seawater with the elements of life (Fig. 1-10).

THE EMERGENCE OF LIFE

Life arose on this planet during a period of crustal formation and volcanic outgassing of an atmosphere and ocean (Table 1-2). It was also a time of heavy meteorite bombardment, and rocky asteroids and icy comets constantly showered the early Earth, possibly providing the main source of the planet's water. Interplanetary space was littered with debris that pounded the newborn planets. Some of this space junk might have provided organic compounds, from which life could evolve. The Earth is still pelted by meteorites that contain amino acids, the precursors of proteins. The meteorite impacts would most likely have made conditions very difficult for proteins to organize into living cells. The first cells might have been repeatedly exterminated, forcing life to originate again and again. Whenever primitive organic molecules attempted to arrange themselves into living matter, frequent impacts blasted them apart before they had a chance to reproduce.

Some large impactors might have generated enough heat to repeatedly evaporate most of the ocean. The vaporized ocean would have raised surface pressures over 100 times greater than the present atmosphere, and the resulting high temperatures would have sterilized the entire planet. Several thousand years would elapse before the steam condensed into rain and the ocean basins refilled again, only to await the next ocean-evaporating impact. Such harsh conditions could have set back the emergence of life hundreds of millions of years.

Perhaps the only safe place for life to evolve was on the deep ocean floor, where a high density of hydrothermal vents existed. Hydrothermal vents are like geysers on the bottom of the ocean that expel mineral-laden hot water heated by shallow magma chambers resting just beneath the ocean floor. The vents might have created an environment capable of generating an immense number of organic reactions. They also would have provided evolving life forms with all the essential nutrients to sustain themselves. Indeed, such an environment exists today, home to some of the strangest creatures found on Earth. In this environment, life could have originated as early as 4.2 billion years ago.

From the very beginning, life had many common characteristics. No matter how varied life is today, from the simplest bacteria to man, its central molecular machinery is exactly the same. Every cell of every organism is constructed from the same set of 20 amino acids. All organisms use the same energy transfer mechanism for growth. All strands of DNA are left-handed double helixes, and the operation of the genetic code in protein synthesis is the same for all living things.

With so much similarity, all life must have sprung from a common ancestor, and all alien forms, of which no descendants exist today, became extinct early in the history of life. Furthermore, no new life forms are being created today either because the present chemical environment is not conducive to the formation of life or living organisms prey upon the newly created life forms before they have a chance to evolve.

Since life appeared within the first half-billion years of the Earth's existence, it must have evolved from simple materials into complex organisms rather quickly. Primitive bacteria, which descended from the earliest known form of life, remain by far the most abundant organisms. Evidence that life began early in the Earth's history, when the planet was still quite hot, exists today as thermophilic (heat-loving) bacteria found in thermal springs and other hot-water environments throughout the world (Fig. 1-11).

The existence of these organisms is evidence that thermophiles were the common ancestors of all life. The early conditions on Earth would have

Figure 1-11   Hot spring water at Yellowstone National Park, Wyoming. Photo by K. E. Barger, courtesy of USGS

been ripe for the evolution of thermophilic organisms, most of which have a sulfur-based energy metabolism; sulfur compounds would have been plentiful on the hot, volcanically active planet.

It is probably fortunate the early Earth had an abundance of sulfur, which was spewed out of a profusion of volcanoes. As long as surface temperatures were hot, ring molecules of sulfur atoms in the atmosphere would block out solar ultraviolet radiation. Otherwise, the first living cells would have sizzled in the deadly rays of the sun. However, an ultraviolet shield might not have been necessary in the primordial atmosphere, because some primitive bacteria appear to tolerate high levels of ultraviolet radiation.

The first living organisms were extremely small noncellular blobs of protoplasm. The self-duplicating organisms fed on a rich broth of organic molecules generated in the primordial sea. Such a nutritional abundance set off a rapid chain reaction, resulting in a phenomenal growth. The organisms drifted freely in the ocean currents and dispersed to all parts of the world. Although the first simple organisms appear to have arrived soon after conditions on Earth became favorable, almost another billion years passed before life even remotely resembled anything living today.

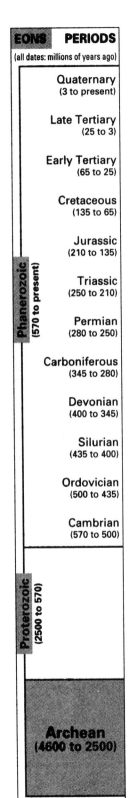

2

ARCHEAN ALGAE

The first 4 billion years of Earth history, or about nine-tenths of geologic time (Table 2-1), are referred to as the Precambrian, the longest and least understood of all eras. The Precambrian is divided nearly equally into the Archean and Proterozoic eons. The boundary is somewhat arbitrary and reflects major differences between the types of rocks formed during the two periods. Archean rocks are products of rapid crustal formation, while Proterozoic rocks are more representative of relatively stable modern geology.

The Archean, from 4.6 to 2.5 billion years ago, covers a time when the Earth was in great turmoil and subjected to extensive meteorite bombardment and intense volcanism. The high internal heat of the newborn planet kept the surface well agitated, destroying any semblance of a crust, which is why the first several hundred million years are absent from the geologic record. During this interval, the planet experienced a restlessness that might have been a major factor in the emergence of life so early in the history of the Earth.

THE AGE OF ALGAE

Life in the Archean consisted mostly of bacteria, unicellular or noncellular algae, and clusters of algae called stromatolites (Fig. 2-1), from the Greek

TABLE 2–1 THE GEOLOGIC TIME SCALE

Era	Period	Epoch	Age (millions of years)	First Life Forms	Geology
		Holocene	0.01		
	Quaternary				
		Pleistocene	3	Man	Ice age
Cenozoic		Pliocene	11	Mastodons	Cascades
		Miocene	26	Saber-toothed tigers	Alps
	Tertiary	Oligocene	37		
		Eocene	54	Whales	
		Paleocene	65	Horses Alligators	Rockies
	Cretaceous		135		
				Birds	Sierra Nevada
Mesozoic	Jurassic		210	Mammals	Atlantic
				Dinosaurs	
	Triassic		250		
	Permian		280	Reptiles	Appalachians
	Pennsylvanian		310		Ice age
	Carboniferous				
Paleozoic	Mississippian		345	Amphibians Insects	Pangaea
				Sharks	
	Devonian		400		
	Silurian		435	Land plants	Laursia
	Ordovician		500	Fish	
	Cambrian		570	Sea plants Shelled animals	Gondwana
			700	Invertebrates	
Proterozoic			2500	Metazoans	
			3500	Earliest life	
Archean			4000		Oldest rocks
			4600		Meteorites

Figure 2-1   Stromatolites of the Missoula group at Glacier National Park, Montana. Photo by R. Rezak, courtesy of USGS

word *stroma*, meaning "stony carpet." The oldest evidence of life includes microfossils, which are the remains of ancient microorganisms, and stromatolites, which are layered structures formed by the accretion of fine sediment grains by colonies of cyanobacteria or primitive blue-green algae living on the ocean floor. Stromatolites, however, are only indirect evidence of early life because they are not the remains of the microorganisms themselves but only the sedimentary structures they built.

Early stromatolite fossils exist in 3.5-billion-year-old sedimentary rocks of the Towers Formation of the Warrawoona group in North Pole, western Australia (see Fig. 3-8). The region was once a tidal inlet, overshadowed by tall volcanoes that erupted ash and lava, which flowed into a shallow sea. Thunderclouds hovered over the peaks, and lightning darted back and forth. Furious winds whipped up high waves that pounded the basaltic cliffs of the coastline. Farther inland, hummocks of black basalt dominated the landscape. The rotten-egg stench of sulfur was pervasive. Frequent downpours fed tidal streams that meandered onto a flat expanse of glistening gray mud before reaching the sea. Elsewhere, scattered shallow pools containing highly saline water periodically evaporated, leaving behind a variety of salts. Often, a floodtide washed across the mud flat, shifting the sediments and replenishing the brine pools.

Although the Archean spans almost half the Earth's history, its rocks represent less than 20 percent of the total area exposed at the surface. Furthermore, all known Precambrian rocks have suffered some heating episode and metamorphism. Unlike most ancient rocks in the 3.5- to 3.8-billion-year range throughout the world, only a few like those of the North Pole sequence have a history of low metamorphic temperatures. Therefore, rocks of this region have remained relatively cool throughout geologic history.

Rocks subjected to the intense heat of the Earth's interior have lost all traces of fossilized life. Even in mildly metamorphosed rocks, the existence of microfossils, which are the preserved cell walls of unicellular microorganisms, is often difficult to prove. Most of these apparent fossils are simple spheres with few surface features, composed of inorganic carbon compounds squeezed into spheroids by the growth of mineral grains deposited around them. But some spheres were linked in pairs or in chains, and were unlikely to have been created simply by inorganic processes.

Associated with the North Pole rocks were cherts, extremely hard siliceous rocks containing microfilaments, which are small, threadlike structures of possible bacterial origin. Similar cherts with microfossils of filamentous bacteria are found at eastern Transvaal, South Africa, dating between 3.2 and 3.3 billion years old. They also exist in 2-billion-year-old

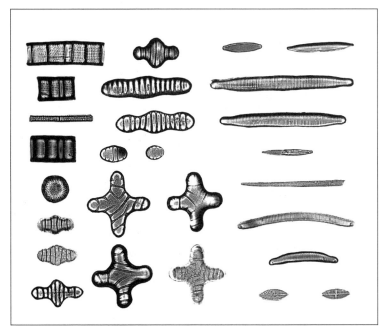

Figure 2-2 Late Miocene diatoms from the Kilgore area, Cherry County, Nebraska. Photo by G. W. Andrews, courtesy of USGS

chert from the Gunflint iron formation on the north shore of Lake Superior in North America. Most Precambrian cherts appear to be chemical sediments precipitated from silica-rich water in deep oceans. The abundance of chert in the Archean might serve as evidence that most of the crust was deeply submerged. However, cherts in the North Pole region appear to have a shallow-water origin.

Chert-forming silica leached out of volcanic rocks that erupted into shallow seas. The silica-rich water circulated through porous sediments, dissolving the original minerals and precipitating silica in their place. Microorganisms buried in the sediments were encased in one of nature's hardest substances, allowing the microfossils to survive the rigors of time. Modern seawater is deficient in silica because organisms like sponges and diatoms (Fig. 2-2) extract it to build their skeletons. Massive deposits of diatomite, also called diatomaceous earth, composed of diatom cell walls are a tribute to the great success of these organisms in the post-Precambrian era.

The North Pole stromatolites are distinctly layered accumulations of calcium carbonate with a rounded, cabbagelike appearance. The size and shape of the Archean-age microfossils and the form of the stromatolites suggest these microorganisms were either oxygen-releasing or sulfur-oxidizing photosynthetic life forms, dependent on sunlight for their growth.

Living stromatolites are similar to those of ancient times and comprise concentrically layered mounds of calcium carbonate built by bacteria or algae, which cement sediment grains together by secreting a jellylike ooze. Older structures at the Australian site are classified either as stromatolite fossils or as layered inorganic sedimentary structures with no biological origin. But microscopic filaments radiating outward from a central point and resembling filamentous (threadlike) bacteria also exist in the fossils, suggesting bacteria built the stromatolites.

Modern stromatolites reside in the intertidal zones above the low tide mark (Fig. 2-3), and their length reflects the height of the tides, which are controlled mostly by the gravitational pull of the moon. The oldest stromatolite colonies of the North Pole region grew to great heights, with some attaining lengths of over 30 feet, which suggests that at an early age the moon orbited much closer to the Earth, and its strong gravitational attraction at this range raised tremendous tides that flooded coastal areas a long distance inland.

The early Earth spun faster on its axis, which meant days were much shorter than today. As the planet's rotation slowed due to drag forces caused by the tides (making days longer), it transferred some of its angular momentum (rotational energy) to the moon, flinging it outward into a widening orbit. Even today, the moon is still receding from the Earth.

Figure 2-3 Algal mounds rising toward the surface from the surrounding lime-stone bottom of a barrier reef in Saipan, Mariana Islands. Photo by P. T. Cloud, courtesy of USGS

THE PROTOZOANS

Simple organisms with primitive cells called prokaryotes, from the Greek word *karyon* meaning "nutshell," lacked a distinct nucleus. They lived under anaerobic (lacking oxygen) conditions and depended mainly on outside sources of nutrients, typically a rich supply of organic molecules continuously created in the sea around them. Most organisms had a primitive form of metabolism called fermentation that converted nutrients into energy. It was an inefficient form of metabolism, releasing energy when enzymes broke down simple sugars such as glucose into smaller molecules.

Each tiny organism was a committee of simpler organelles that the organism incorporated into its cells in a symbiotic relationship, creating a new type of organism called a eukaryote, which was equipped with a nucleus that organized genetic material. When the cell divided, DNA in the nucleus and in the organelles replicated, with half the genes remaining with the parent and the other half passed on to the daughter cell.

This process, called mitosis, increased the likelihood of genetic variation and greatly accelerated the rate of evolution as organisms encountered new environments, and were able to adapt as needed. The extraordinary variety of plant and animal life that has arisen on this planet over the last 600 million years is due exclusively to the introduction of the eukaryotic cell and its huge potential for genetic diversity.

Early single-cell animals called protistids (Fig. 2-4) shared many characteristics with plants. The cells contained elongated structures of mitochondria, which are bacterialike bodies that produce energy by oxidation. They also contained chloroplasts, which are packets of chlorophyll that provide energy by photosynthesis. Many protozoans secreted a tiny shell composed of calcium carbonate. When the animals died, their shells sank to the bottom of the ocean, where over time they built up impressive formations of limestone.

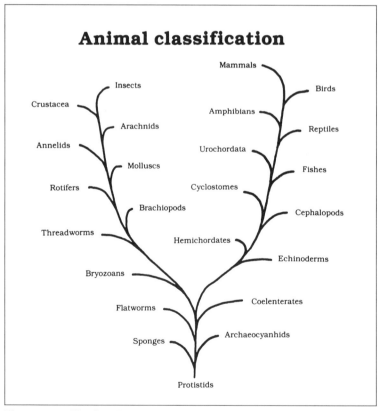

Animal classification

Figure 2-4 The family tree of animals.

Figure 2-5 Tube worms on the East Pacific Rise. Courtesy of Woods Hole Oceanographic Institution

The ability to move about under their own power is what essentially separates animals from plants, although some animals perform this function only in the larval stage and become sedentary or fixed to the seabed as adults. Mobility enabled animals to feed on plants and other animals, establishing a new predator-prey relationship.

Some organisms moved about by a thrashing tail called a flagellum, resembling a filamentous bacterium that joined the host cell for mutual benefit. Other cells had tiny hairlike appendages called cilia that propelled the organism around by rhythmically beating the water. Many, like the amoeba, traveled by extending fingerlike protrusions outward from the main body and flowing into them.

The earliest organisms were sulfur-metabolizing bacteria similar to those living symbiotically in the tissues of tube worms (Fig. 2-5), which live near sulfurous hydrothermal vents on the East Pacific Rise and also on the Gorda Ridge off the northwest Pacific coast of the United States. Sulfur was particularly abundant in the early ocean and combined easily with metals like iron to form sulfates. Since the atmosphere and ocean lacked oxygen, the bacteria obtained energy by the reduction of sulfate ions. The growth of primitive bacteria was thus limited by the amount of organic molecules produced in the ocean. Although this form of energy was satisfactory, bacteria were letting a plentiful source of energy go to waste, namely sunlight.

PHOTOSYNTHESIS

The ratios of carbon isotopes in Archean rocks suggest that photosynthesis was in progress at an early age. The seas contained an abundance of iron, and oxygen generated by photosynthesis was lost by oxidation with this element, a fortunate circumstance since oxygen was also poisonous to primitive life forms. Abundant sulfur in the early sea provided the nutrients to sustain life without the need for oxygen, and bacteria obtained energy by the reduction (opposite of oxidation) of this important element.

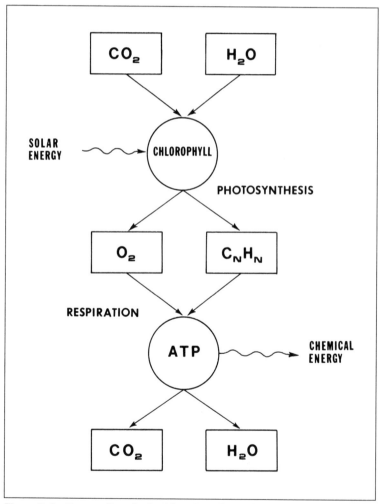

Figure 2-6 Energy flow in the biosphere from the input of solar energy to the utilization of this energy by photosynthesis and respiration.

A primitive form of photosynthesis probably began about 3.5 billion years ago with the first appearance of blue-green algae, or its predecessor, a photosynthetic bacteria called green sulfur bacteria. These organisms were best suited to an oxygen-poor environment. Oxygen was kept to a minimum by reacting with both dissolved metals in seawater and gases emitted from submarine hydrothermal vents.

The first green-plant photosynthesizers called proalgae were probably intermediate between bacteria and blue-green algae and could switch from fermentation, a primitive form of metabolism, to photosynthesis and back again, depending on their environment. Since sunlight penetrates seawater to a maximum effective depth of about 300 feet, the proalgae were confined to shallow water. Around 2.8 billion years ago, microorganisms called cyanobacteria began to use sunlight as their main energy source to drive the chemical reactions needed for sustained growth.

The development of photosynthesis was possibly the single most important step in the evolution of life. It gave a primitive form of blue-green algae a practically unlimited source of energy. Photosynthesis utilized sunlight to split water molecules, and the hydrogen combined with carbon dioxide that was abundant in the early ocean and atmosphere to form simple sugars and proteins, liberating oxygen in the process (Fig. 2-6). The growth of photosynthetic organisms was phenomenal, and the population explosion would have gotten out of hand except that oxygen, generated as a waste product of photosynthesis, was also poisonous. If not for the development

TABLE 2–2 EVOLUTION OF LIFE AND THE ATMOSPHERE

Evolution	Origin (million years)	Atmosphere
Humans	2	Nitrogen, oxygen
Mammals	200	Nitrogen, oxygen
Land animals	350	Nitrogen, oxygen
Land plants	400	Nitrogen, oxygen
Metazoans	700	Nitrogen, oxygen
Sexual reproduction	1100	Nitrogen, oxygen, carbon dioxide
Eukaryotic cells	1400	Nitrogen, carbon dioxide, oxygen
Photosynthesis	2300	Nitrogen, carbon dioxide, oxygen
Origins of life	3800	Nitrogen, methane, carbon dioxide
Origins of Earth	4600	Hydrogen, helium

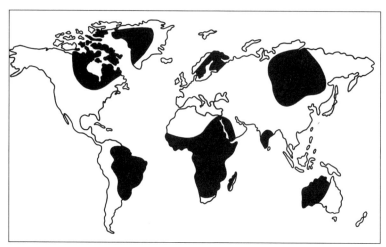

Figure 2-7   The location of continental shields, the foundations upon which the continents grew.

of special enzymes to help organisms cope with and later use oxygen for their metabolism, life would certainly have been in jeopardy.

To create and maintain an oxygen-rich atmosphere, the carbon dioxide used in the photosynthetic process had to be buried as carbonate rock faster than the oxygen was consumed by the oxidation of carbon, metals, and volcanic gases. About 2 billion years ago, oxygen began replacing carbon dioxide in the ocean and atmosphere. Therefore, organisms had to either develop a means of shielding their nuclei and other critical sites from oxygen or use chemical pathways that removed hydrogen instead of adding oxygen. These innovations led to the evolution of the eukaryotes. In this manner, oxygen was responsible in large part for the evolution of higher forms of life (Table 2-2).

GREENSTONE BELTS

Greenstones are ancient metamorphosed rocks unique to the Archean. About 4 billion years ago, the mantle cooled enough to form a permanent crust composed of a thin layer of basalt embedded with scattered blocks of granite called "rockbergs." The granite combined into stable bodies of basement rock, upon which all other rocks were deposited. The basement rocks became the nuclei of the continents and are presently exposed in broad, low-lying, domelike structures called shields (Fig. 2-7).

The Precambrian shields are extensive uplifted areas surrounded by sediment-covered bedrock called continental platforms, which are broad, shallow depressions of basement complex (crystalline rock) filled with nearly flat-lying sedimentary rocks. The best-known areas are the Canadian

Shield in North America and the Fennoscandian Shield in Europe. More than a third of Australia is Precambrian shield, and sizable shields exist in Africa, South America, and Asia. Many shields are fully exposed where flowing ice sheets eroded their cover of sediment during the last ice age.

Dispersed among and around the shields are greenstone belts, which are a mixture of metamorphosed (recrystallized) lava flows and sediments possibly derived from island arcs (chains of volcanic islands) caught between colliding continents. Although no large continents existed during this time, the foundations upon which they formed were present as proto-

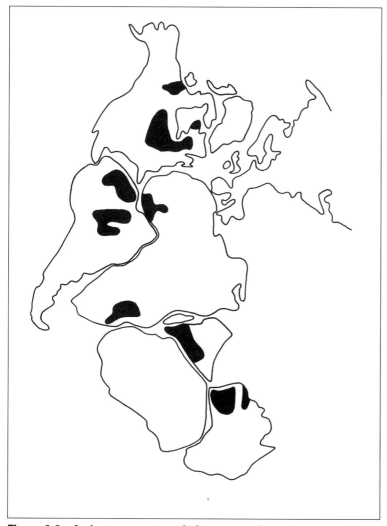

Figure 2-8 Archean greenstone belts occupy the ancient cores of the continents.

Figure 2-9　Pillow lava on the south bank of Webber Creek, Eagle District, Alaska.
Photo by E. Blackwelder, courtesy of USGS

continents. These small landmasses were separated by marine basins that accumulated lava and sediments derived mainly from volcanic rocks that later recrystallized into greenstone belts.

Greenstone belts occupy the ancient cores of the continents (Fig. 2-8). They span an area of several hundred square miles, surrounded by immense expanses of gneisses, which are the metamorphic equivalents of granites and the predominant Archean rock types. Greenstone is green because of the mineral chlorite, a greenish form of mica. The best-known greenstone belt is the Swaziland sequence in the Barberton Mountain Land of southeastern Africa. It is over 3 billion years old and is at points nearly 12 miles thick.

Caught in the Archean greenstone belts are ophiolites, the name derived from the Greek word *ophis,* meaning "serpent." They are slices of ocean floor shoved up on the continents by drifting plates and are as much as 3.6 billion years old. Pillow lavas (Fig. 2-9), which are tubular bodies of basalt extruded undersea, also appear in the greenstone belts, signifying that the volcanic eruptions took place on the ocean floor. Thus, these deposits are among the best evidence that plate tectonics operated early in the Archean. Indeed, plate tectonics appears to have been working throughout most of the Earth's history in much the same manner as it does today.

The greenstone belts are of particular interest to geologists (and miners) because they hold most of the world's gold deposits. Archean-age ore deposits are remarkably similar worldwide. The mineralized veins are either Archean in age or they invaded Archean rocks at a much later date. Gold of Archean age is mined on every continent except Antarctica. In

Africa, the best gold deposits are in rocks as much as 3.4 billion years old. Most of the South African gold mines are in greenstone belts.

The Kolar greenstone belt in India, formed when two plates crashed together about 2.5 billion years ago, holds the richest gold deposits in the world. In North America, the best gold mines are in the Great Slave region of northwest Canada, where over a thousand deposits are known. Most of the gold deposits exist in greenstone belts invaded by hot magmatic solutions originating from the intrusion of granitic bodies into the green-stones, and the gold occurs in veins associated with quartz.

Because greenstone belts have no equivalent in modern geology, the geologic conditions under which they formed were very different from those observed today. Active tectonic (landform-building) forces in the mantle often broke open the thin Archean crust and injected magma into the deep crustal fracture zones. Such large-scale magmatic intrusion along

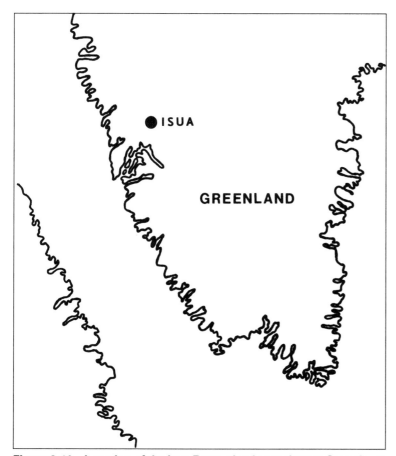

Figure 2-10   Location of the Isua Formation in southwest Greenland, where some of the world's oldest rocks lie.

with numerous large meteorite impacts characterized the unusual geology of the Archean. Since greenstone belts are geologically unique to the Archean, their absence in geologic formations after 2.5 billion years ago marks the end of the eon.

ARCHEAN CRATONS

Plate tectonics has played a prominent role in shaping the planet practically from the very beginning. Continents were adrift from the time they originated, within a few hundred million years after the formation of the planet. This is revealed by the presence of 4-billion-year-old Acasta gneiss, a metamorphosed granite, in the Northwest Territories of Canada, which suggests that the formation of the crust was well underway by this time.

The discovery leaves little doubt that at least small patches of continental crust existed on the Earth's surface during the first half-billion years of its history. The 3.8-billion-year-old metamorphosed marine sediments of the Isua Formation in a remote mountainous area in southwest Greenland (Fig. 2-10) provide evidence of an ocean. The continental crust was perhaps only about 10 percent of its present size and contained slivers of granite that drifted freely over the Earth's watery face.

Few rocks date beyond 3.7 billion years, suggesting that little continental crust was formed until afterward or was recycled into the mantle. Slices of granitic crust combined into stable bodies of basement rock called cratons (Fig. 2-11), upon which all other rocks were deposited. They are composed of highly altered granite and metamorphosed marine sediments and lava flows. The rocks originated from intrusions of magma into the primitive ocean crust. Only three sites in the world, in Canada, Australia, and Africa, contain rocks that

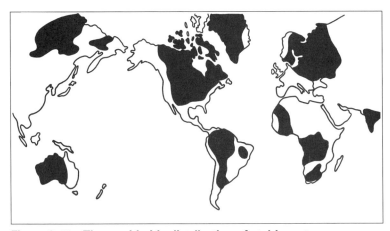

Figure 2-11   The worldwide distribution of stable cratons.

were exposed on the surface during the Earth's early history and have remained essentially unchanged throughout geologic time.

Eventually, the slices of crust began to slow their erratic wanderings and combine into larger landmasses. Constant bumps and grinds from vigorous tectonic activity built the crust both inside and out. The continents continued growing rapidly until the end of the Archean 2.5 billion years ago, when they occupied up to a quarter of the Earth's surface, or about 80 percent or more of the present continental landmass. During this time, plate tectonics began to operate extensively, and much of the world as we know it took shape.

Ophiolites are the best evidence for ancient plate motions. They are sections of oceanic crust that peeled off during plate collisions and were plastered onto the continents. Blueschists (Fig. 2-12), which are metamorphosed rocks of subducted ocean crust, also shoved up on the continents. This resulted in a linear formation of greenish volcanic rocks along with light-colored masses of granite and gneiss, which are common igneous and metamorphic rocks that comprise the bulk of the continents.

The cratons numbered in the dozens and ranged from about a fifth the area of present-day North America to smaller than the state of Texas. The cratons were highly mobile and moved about freely on the molten rocks of the upper mantle called the asthenosphere, becoming independent minicontinents that periodically collided with and rebounded off each other. The collisions crumpled the leading edges of the cratons, forming small parallel mountain ranges perhaps only a few hundred feet high.

Figure 2-12 An outcrop of retrograde blueschist rocks, evidence of plate tectonics, in the Seward Peninsula region, Alaska. Photo by C. L. Sainsbury, courtesy of USGS

All the cratons eventually coalesced into a single large landmass several thousands of miles wide called a supercontinent. The points at which the cratons collided saw mountain ranges forced up, and the sutures joining the landmasses are still visible today as cores of ancient mountains called orogens. The original cratons formed within the first 1.5 billion years of the Earth's existence and totaled only about a tenth of the present landmass. The average rate of continental growth since then has been perhaps as much as one cubic mile a year. The constant rifting and patching of the interior along with sediments deposited along the continental margins eventually built the supercontinent outward so that by the end of the Archean it nearly equaled the total area of today's continents.

3

PROTEROZOIC METAZOANS

EONS	PERIODS
	(all dates: millions of years ago)

EONS	PERIODS (all dates: millions of years ago)
Phanerozoic (570 to present)	Quaternary (3 to present)
	Late Tertiary (25 to 3)
	Early Tertiary (65 to 25)
	Cretaceous (135 to 65)
	Jurassic (210 to 135)
	Triassic (250 to 210)
	Permian (280 to 250)
	Carboniferous (345 to 280)
	Devonian (400 to 345)
	Silurian (435 to 400)
	Ordovician (500 to 435)
	Cambrian (570 to 500)
	Proterozoic (2500 to 570)
Archean (4600 to 2500)	

Major differences exist in the character of rocks of the Proterozoic eon, from 2.5 to 0.6 billion years ago, as compared to those of the Archean. The Proterozoic featured a shift to much calmer conditions, as the Earth progressed from a tumultuous adolescence to a stable adulthood. Marine life was distinct from that of the Archean and represented a considerable advancement in evolution with the development of complex organisms.

The global climate was cooler, and the Earth experienced its first major ice age more than 2 billion years ago, along with a major extinction that eliminated many primitive species attempting to evolve during this time. The Proterozoic ended about 570 million years ago after a second period of glaciation and another mass extinction. Afterward, an explosion of species, representing nearly every major group of marine organisms, set the stage for the evolution of more modern forms of life.

THE AGE OF WORMS

Life in the Proterozoic was more advanced and complex than in the Archean. Organisms evolved very little during their first billion years on

Earth because of primitive, asexual reproduction, which used simple fission, whereby species cloned themselves, offering little evolutionary change. A primitive form of metabolism also kept the organisms in a low-energy state.

The first major evolutionary advancements were the development of an organized nucleus and sexual reproduction, which introduced a new breed of single-celled organisms called eukaryotes, around 1.5 billion years ago. Metabolism in eukaryotes involves respiration, indicating that substantial quantities of oxygen were present by the Proterozoic. About 2 billion years ago, when the banded iron formations, which absorbed oxygen, were no longer being deposited, oxygen began to replace carbon dioxide in the ocean and atmosphere.

About 1.5 billion years ago, the previously skimpy geologic record of preserved cellular remains vastly improves as evolution suddenly sped up.

TABLE 3–1   THE CLASSIFICATION OF SPECIES

Group	Characteristics	Geologic Age
Vertebrates	Spinal column and internal skeleton. About 70,000 living species. Fish, amphibians, reptiles, birds, mammals.	Ordovician to recent
Echinoderms	Bottom dwellers with radial symmetry. About 5,000 living species. Starfish, sea cucumbers, sand dollars, crinoids.	Cambrian to recent
Arthropods	Largest phylum of living species with over 1 million known. Insects, spiders, shrimp, lobsters, crabs, trilobites.	Cambrian to recent
Annelids	Segmented body with well-developed internal organs. About 7,000 living species. Worms and leeches.	Cambrian to recent
Mollusks	Straight, curled, or two symmetrical shells. About 70,000 living species. Snails, clams, squids, ammonites.	Cambrian to recent
Brachiopods	Two asymmetrical shells. About 120 living species.	Cambrian to recent
Bryozoans	Moss animals. About 3,000 living species.	Ordovician to recent
Coelenterates	Tissues composed of three layers of cells. About 10,000 living species. Jellyfish, hydra, coral.	Cambrian to recent
Porifera	The sponges, about 3,000 living species.	Proterozoic to recent
Protozoans	Single-celled animals. Forams and radiolarians.	Precambiran to recent

However, another three-quarters of a billion years elapsed before multicellular animals called metazoans appear in the fossil record. By then, the dissolved oxygen content of the sea was about 5 to 10 percent of its present value. The increased level of oxygen also appears to have sparked the evolution of many unique animals.

The triggering mechanisms for such a rapid evolutionary phase included ecological stress, geographic isolation caused by drifting continents, and climatic changes. Organisms no longer relied entirely on surface absorption of oxygen, and gills and circulatory systems evolved when the oxygen level reached about a tenth of its present value near the end of the Proterozoic. Afterward, an explosion of species created the progenitors of all life on Earth today (Table 3-1).

The first metazoans were a loose organization of individual cells united for a common purpose, such as locomotion, feeding, or protection. The

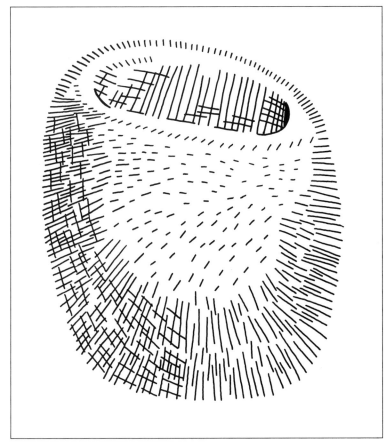

Figure 3-1 Early sponges were among the first giants on the ocean floor.

most primitive metazoans probably comprised numerous cells, each with its own flagellum. They grouped into a small, hollow sphere, and their flagella beat the water in unison to propel the tiny animal through the sea.

From these metazoans evolved sedentary forms turned inside out and attached to the ocean floor. Openings to the outside enabled the flagella now on the inside to produce a flow of water, providing a crude circulatory system for filtering food particles and ejecting wastes. These were the forerunners of the sponges (Fig. 3-1), the most primitive of metazoans. They existed in various shapes and sizes, some species possibly reaching 10 feet or more across, and grew in thickets on the ocean floor. The body consisted of three weak tissue layers, whose cells could survive independently if separated from the main body. The cells could either reattach or grow separately into mature sponges.

The next evolutionary step was the jellyfish, which had two layers of cells separated by a gelatinous substance, giving the saucerlike body a means of support. Unlike the sponges, the cells of the jellyfish were incapable of independent survival if separated from the main body. A primitive nervous system linked the cells and caused them to contract in unison, thereby providing the first simple muscles used for locomotion.

The development of muscles and other rudimentary organs, including sense organs and a central nervous system to process the information, followed the evolution of primitive segmented worms. The coelomic, or

Figure 3-2 Fossil worm borings in Heiser Sandstone, Pensacola Mountains, Antarctica. Photo by D. L. Schmidt, courtesy of USGS

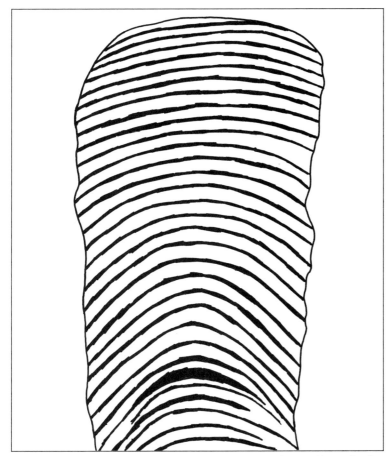

Figure 3-3   Stromatolites are layered structures formed by colonies of primitive blue-green algae.

hollow-bodied, worms adapted to burrowing in the ocean floor sediments and might have evolved into higher forms of animal life. Since they were bottom-dwellers, these early worms left behind a preponderance of fossilized tracks, trails, and burrows (Fig. 3-2) to such an extent that the Proterozoic is often referred to as the "age of worms." Prior to about 670 million years ago, however, no track-making animals existed.

Sheetlike marine worms were less than a tenth of an inch thick but grew nearly three feet long, providing a large surface area on which to absorb oxygen and nutrients directly from seawater. Another reason for the unusual flattened bodies of many animals was the high ratio of surface area to volume it created, a result of the limited food supply available during the Proterozoic. This high ratio allowed for more effective sunlight collection for algae, which lived within the bodies of worms, helping to nourish their hosts while the worms supported them in symbiosis.

Algae also built tall stromatolite structures (Fig. 3-3), composed of concentric layers of sediment, that tilted in the average direction of the sun. Stromatolite fossils in the Bitter Springs Formation in central Australia provide an 850-million-year-old record of the sun's movement across the sky. A stromatolite mound situated near the equator pointed toward one pole in the winter and the other in the summer, developing a sinuous growth pattern in the form of an S.

If a stromatolite laid down sediment layers daily, the number of layers in one full sine wave represented the number of days in a year. Analysis of stromatolites living in the late Proterozoic indicate an estimated 435 days in a single year. The results agree with counts made of growth rings of ancient coral fossils dating back to the beginning of the Cambrian period, 570 million years ago. Then, a year was about 428 days; 200 million years later, in the middle of the Devonian period, the figure became 400 days (Fig. 3-4). The longer year indicates a higher rotation rate for the planet, due to the proximity of the moon, which was perhaps half its present distance from the Earth at the time it was created in the early Archean. Tidal friction gradually slowed the Earth's rotation and flung the moon into a higher orbit.

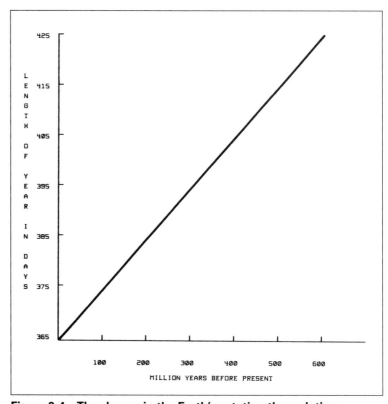

Figure 3-4   The change in the Earth's rotation through time.

In addition, the sine wave patterns of ancient stromatolites provide information about the maximum travel of the sun across the equator. The equator forms an oblique angle to the ecliptic, controlled by the tilt of the Earth's rotational axis. The sun's maximum latitude during the peak of each season is obtained by measuring the maximum angle the sine wave deviates from the average direction of stromatolite growth. Presently, the sun travels 23.5 degrees either side of the equator from summer to winter. However, around 850 million years ago, this value was about 26.5 degrees, suggesting a much more seasonal climate.

THE EDIACARA FAUNA

During the late Proterozoic, around 800 million years ago, stromatolites underwent a marked decline in diversity possibly due to the appearance of algae-eating animals. Around 680 million years ago, thick glaciers spread over the continents during perhaps the most intense period of glaciation the Earth has ever known, when massive ice sheets overran nearly half the land surface. At this time, all continents were assembled into a supercontinent, which might have wandered over one of the poles. The ice age dealt a deathly blow to life in the ocean, and many simple organisms vanished during the world's first mass extinction. At this point in the Earth's development, animal life was still scarce, and the extinction decimated the ocean's population of acritarchs, a community of planktonic algae that were among the first organisms to develop elaborate cells with nuclei.

Not long after the ice disappeared near the end of the Proterozoic, about 670 million years ago, the great diversity of animal life culminated with the evolution of entirely new species, the likes of which had never existed before, or since, forever changing the Earth's biology. Life forms took off in all directions, producing many unique and bizarre creatures, whose fossil impressions are found in the Ediacara formation in South Australia. But many soft-bodied organisms like the late Precambrian fauna living prior to the arrival of shelled animals, did not easily enter the fossil record, another reason only a small fraction of all species that have ever lived are preserved as fossils.

The extremely flattened bodies of the Ediacaran fauna maximized the ratio of surface area to volume, enabling organisms to take in nutrients and oxygen more efficiently and to absorb light for symbiotic algae. The algae lived within the tissues of host animals, which offered protection from predators in return for nutrients and the removal of waste products. These adaptions served well for the marine conditions that prevailed in the late Proterozoic, when shallow seas were nutrient poor and oxygen levels were low.

The Earth underwent many profound physical changes near the end of the Proterozoic, prompting a rapid radiation of Ediacaran fauna. At the same time, a supercontinent located on the equator rifted apart, producing

Figure 3-5   The archaeocyathans built the earliest limestone reefs.

intense hydrothermal activity that caused fundamental environmental changes. Furthermore, the increased marine habitat area spawned the greatest explosion of new species the world has ever known, with seas that contained large populations of widespread and diverse organisms.

The dominant animals were the coelenterates, radially symmetrical invertebrate animals, including giant jellyfishlike floaters up to 3 feet wide and colonies of feathery forms, possibly predecessors of the corals, that were attached to the ocean floor and grew more than a yard long. The remaining organisms were mostly marine worms, unusual arthropodlike animals, and a tiny, curious looking, naked starfish with three rays instead of the customary five. The vase-shaped archaeocyathans (Fig. 3-5) resembled both sponges and corals and built the earliest limestone reefs, eventually becoming extinct in the Cambrian.

The Ediacara formation contains impressions of strange organisms. Many of these unusual creatures were the result of adaptations to highly unstable conditions during the late Proterozoic. These included an increasing oxygen supply, which made possible the evolution of large animals with vascular circulatory systems supplying the cells with blood. Over-specialization to a narrow range of environmental conditions caused a major extinction of species at the end of the era around 570 million years ago due to the evolution of predators. Those species that survived the great extinction were markedly different from their Ediacaran ancestors.

BANDED IRON FORMATIONS

Mineral deposits of the Proterozoic are bedded in layers, or stratified, as opposed to the vein deposits of the Archean. Iron, the fourth most abundant element in the Earth's crust, was leached from the continents and dissolved in seawater under reducing (deoxydizing) conditions. When the iron reacted with oxygen in the ocean, it precipitated in vast deposits on shallow continental margins. Alternating bands of iron-rich and iron-poor sediments gave the ore a banded appearance, thus prompting the name banded iron formation, or BIF. These deposits, mined extensively throughout the world, provide over 90 percent of the minable iron reserves.

In effect, biologic activity was responsible for the iron deposits, since photosynthetic organisms produced the oxygen. When plants began producing oxygen (Fig. 3-6), it combined with iron, keeping oxygen levels in the ocean within the limits tolerated by the early prokaryotes. Throughout the Archean, the amount of oxygen probably remained under 1 percent due to this regulating mechanism. Then, between 2.5 and 2 billion years ago, photosynthesis generated enough oxygen to react with iron on a grand scale.

BIF deposits, composed of alternating layers of iron and silica, formed about 2 billion years ago at the height of the earliest ice age. For unknown reasons, major episodes of iron deposition coincided with periods of glaciation. The average ocean temperature was probably warmer than today. When warm ocean currents rich in iron and silica flowed toward the glaciated polar regions, the suddenly cooled waters could no longer hold minerals in solution. As they became "undissolved" in the water, their precipitation formed alternating layers on the ocean floor, alternating because iron, heavier than silica, settles faster. After most of the dissolved iron was locked up in the sediments, the level of oxygen began to steadily rise, spawning the evolution of more advanced species.

Biochemical activity in the ocean was also responsible for stratified sulfide deposits. Sulfur-metabolizing bacteria living near undersea hydrothermal vents oxidized hydrogen sulfide into elemental sulfur and

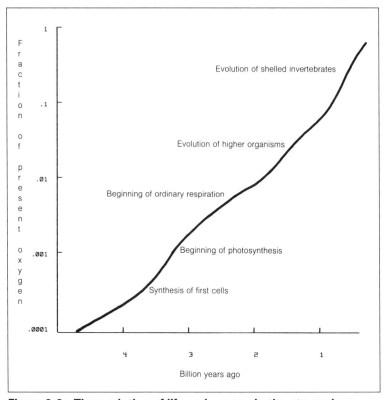

Figure 3-6 The evolution of life and oxygen in the atmosphere.

various sulfates. Copper, lead, and zinc, which were much more abundant in the Proterozoic than in the Archean, also reflect a submarine volcanic origin.

PRECAMBRIAN GLACIATION

The Proterozoic was a period of transition, when oxygen generated by photosynthesis replaced carbon dioxide. Early in the Archean, the sun's output was only about 70 percent of its present value and large amounts of atmospheric carbon dioxide, 1,000 times greater than current levels, kept the Earth's oceans from freezing solid.

When the first microscopic plants evolved, they began replacing carbon dioxide in the ocean and atmosphere with oxygen so that today the relative percentages of these two gases have completely reversed. The loss of carbon dioxide, an important greenhouse gas, caused the climate to cool, even while the sun was getting progressively hotter.

The drop in global temperatures soon after the beginning of the Proterozoic initiated the first known glacial epoch about 2.4 billion years ago
(Table 3-2), when massive sheets of ice nearly engulfed the entire landmass.
The positions of the continents also had a tremendous influence on the
initiation of ice ages, and landmasses wandering into the colder latitudes
enabled the buildup of glacial ice. Global tectonics, featuring extensive
volcanic activity and seafloor spreading, might have triggered the ice ages
by drawing down the level of oxygen in the ocean and atmosphere,
preserving more organic carbon in the sediments so that living organisms
could not return it to the atmosphere.

Plate tectonics also began to operate more vigorously, and carbonaceous
sediments were thrust deep inside the Earth along with the underlying
oceanic crust. The growing continents stored large quantities of carbon in
thick deposits of carbonaceous rocks such as limestone. The elimination of
carbon dioxide in this manner caused the Earth to cool dramatically.
Besides high rates of organic carbon burial, the iron deposition and intense
hydrothermal activity associated with plate tectonics furthered global
cooling. Although this was the first ice age the world had ever known, it
was not the worst.

The burial of carbon in the Earth's crust might have been the key to the
onset of another glacial period near the end of the Proterozoic about 680
million years ago. A supercontinent located on the equator rifted apart,

TABLE 3–2   THE MAJOR ICE AGES

Time (in years)	Event
10,000–present	Present interglacial
15,000–10,000	Melting of ice sheets
20,000–18,000	Last glacial maximum
100,000	Most recent glacial episode
1 million	First major interglacial
2 million	First glacial episode in Northern Hemisphere
4 million	Ice covers Greenland and the Arctic Ocean
15 million	Second major glacial episode in Antarctica
30 million	First major glacial episode in Antarctica
65 million	Climate deteriorates, poles become much colder
250–65 million	Interval of warm and relatively uniform climate
250 million	The great Permian ice age
700 million	The great Precambrian ice age
2.4 billion	First major ice age

forming four or five major continents, the largest of which apparently wandered into the south polar region and acquired a thick blanket of ice. This was perhaps the greatest period of glaciation, and ice encased nearly half the landmass. The climate was so cold ice sheets and permafrost (permanently frozen ground) extended toward the equator. During this time, no plants grew on the land and only simple plants and animals lived in the sea.

The glacial periods are marked by deposits of tillites. Thick sequences of Precambrian tillites exist on every continent (Fig. 3-7). Tillites are a mixture of boulders and clay deposited by glacial ice and cemented into solid rock. In the Lake Superior region of North America, tillites are 600 feet thick in places and range from east to west for a thousand miles. In northern Utah, tillites mount up to 12,000 feet thick. The various layers of glacially deposited sediment suggest a series of ice ages closely following each other. Similar tillites are among Precambrian rocks in Norway, Greenland, China, India, southwest Africa, and Australia.

The 680-million-year-old glacial varves in lakebed deposits north of Adelaide, South Australia might hold evidence of a solar cycle operating in the Proterozoic. The varves consist of alternating layers of silt laid down annually during the late Precambrian ice age (Fig. 3-8). Each summer when the glacial ice melted, sediment-laden meltwater discharged into a lake, and sediments settled out in a stratified deposit.

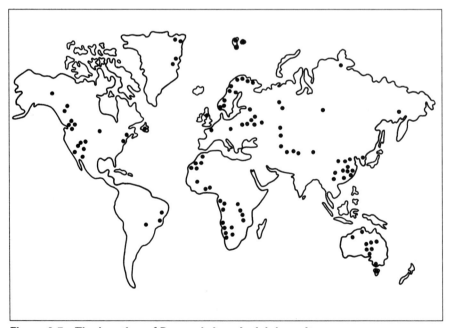

Figure 3-7   The location of Precambrian glacial deposits.

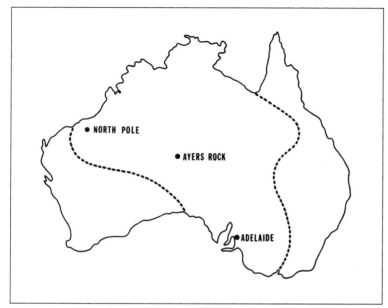

Figure 3-8   The extent of the late Precambrian ice age in Australia.

During times of intense solar activity, average global temperatures increased, and the glaciers melted more rapidly, depositing thicker varves. By counting the layers of thick and thin varves, a stratigraphic sequence is established that mimics both the 11-year sunspot cycle and the 22-year solar cycle or possibly the early lunar cycle, which today is about 19 years.

When the glacial epoch ended and the ice sheets retreated, life began to proliferate in the ocean with an intensity never experienced before or since. The rapid evolution produced three times as many phyla, groups of organisms sharing the same general body plan, as are living today, many of them unique and bizarre creatures.

THE CONTINENTAL CRUST

By the beginning of the Proterozoic, upwards of 80 percent or more of the present continental crust was already in existence (Table 3-3). Most of the material presently locked up in sedimentary rocks was at or near the surface, and ample sources of Archean rocks were exposed to erosion and redeposition. Sediments derived directly from primary sources are called graywackes, often described as dirty sandstone and common in folded sedimentary rocks such as those in the Alps and Alaska. Most Proterozoic wackes composed of sandstones and siltstones originated from Archean greenstones. Another common rock-type was fine-grained quartzite, a

TABLE 3–3   CLASSIFICATION OF THE EARTH'S CRUST

Environment	Crust Type	Tectonic Character	Thickness (in miles)	Geologic Features
Continental crust overlying stable mantle	Shield	Very stable	22	Little or no sediment, exposed Precambrian rocks
	Midcontinent	Stable	24	
	Basin and Range	Very unstable	20	Recent normal faulting, volcanism, and intrusion; high mean elevation
Continental crust overlying unstable mantle	Alpine	Very unstable	34	Rapid recent uplift, relatively recent intrusion; high mean elevation
	Island arc	Very unstable	20	High volcanism, intense folding, and faulting
Oceanic crust overlying stable mantle	Ocean basin	Very stable	7	Very thin sediments overlying basalts, no thick Palaeozoic sediments
Oceanic crust overlying unstable mantle	Ocean ridge	Unstable	6	Active basaltic volcanism, little or no sediment

Figure 3-9   Ladore Canyon, looking north toward Browns Park, Uinta Mountains, Summit County, Utah. Photo by W. R. Hansen, courtesy of USGS

metamorphic rock derived from the erosion of siliceous grainy rocks such as granite and a coarse-grained sandstone called arkose.

Conglomerates, consolidated gravel-like deposits, were also abundant in the Proterozoic. Nearly 20,000 feet of Proterozoic sediments form the Uinta Range of Utah (Fig. 3-9), and the Montana Proterozoic belt system contains sediments over 11 miles thick. The Proterozoic is also known for its widespread terrestrial redbeds, named so because sediment grains were cemented with red iron oxide. Their appearance around 1 billion years ago indicates that the atmosphere had substantial levels of oxygen at this time.

The weathering of primary rocks produced solutions of calcium carbonate, magnesium carbonate, calcium sulfate, and sodium chloride, which subsequently precipitated into limestone, dolomite, gypsum, and halite. Higher carbon dioxide concentrations in the Precambrian probably account

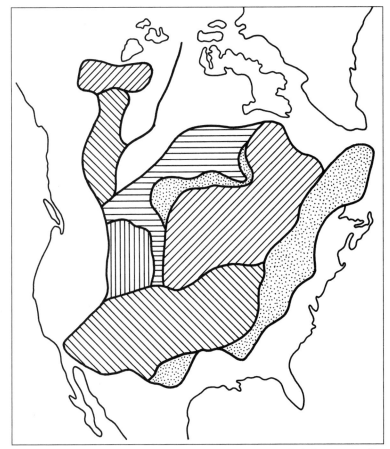

Figure 3-10   The cratons (continental building blocks) that comprise the North American continent.

for the prevalence of dolomite over limestone. These minerals appear to be mainly chemical precipitates and not of biological origin.

In the Mackenzie Mountains of northwest Canada, dolomite deposits range up to 6,500 feet thick, and in the great Alps, massive chunks of dolomite soar skyward. Carbonate rocks such as limestone and chalk, produced chiefly by an organic process involving shells and skeletons of simple organisms, became more common during the late Proterozoic between about 700 and 570 million years ago. In contrast, they were relatively rare in the Archean due to the scarcity of lime-secreting organisms.

The continents of the Proterozoic were composed of Archean cratons. The North American continent assembled from seven cratons around 2 billion years ago (Fig. 3-10), making it one of the oldest continents. The cratons welded into what is now central Canada and the north-central United States. At Cape Smith on Hudson Bay is a 2-billion-year-old slice of oceanic crust that was accreted onto the land, indicating that continents collided and closed an ancient ocean. Arcs of volcanic rock also weave through central and eastern Canada down into the Dakotas. A region between Canada's Great Bear Lake and the Beaufort Sea holds the roots of an ancient mountain range that formed by the collision of North America and an unknown landmass between 1.2 and 0.9 billion years ago. Meanwhile, continental collisions continued to add a large area of new crust to the growing proto-North American continent.

Most of the continental crust underlying the United States from Arizona to the Great Lakes to Alabama formed in one great surge of crustal generation unequaled in North America since. The rapid buildup of new crust possibly resulted from the most energetic period of tectonic activity in Earth history. The assembled North American continent was stable enough to resist a billion years of jostling and rifting, and continued to grow by adding bits and pieces of continents and island arcs to its edges.

Large masses of volcanic rock found near the eastern edge of North America imply that the continent was the core of a larger supercontinent formed during the late Proterozoic. The interior of this landmass heated and erupted with volcanism. The warm, weakened crust consequently broke apart into possibly four or five major continents between 630 and 560 million years ago, although they were shaped differently than they are today. The breakup of the supercontinent created thousands of miles of new continental margin, which might have played a major role in the explosion of new life at the beginning of the Phanerozoic.

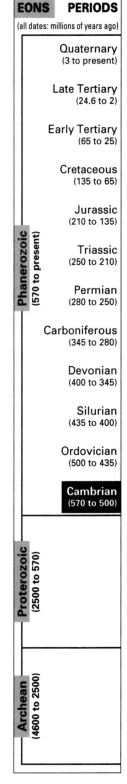

4

CAMBRIAN INVERTEBRATES

Quaternary
(3 to present)

Late Tertiary
(24.6 to 2)

Early Tertiary
(65 to 25)

Cretaceous
(135 to 65)

Jurassic
(210 to 135)

Phanerozoic
(570 to present)

Triassic
(250 to 210)

Permian
(280 to 250)

Carboniferous
(345 to 280)

Devonian
(400 to 345)

Silurian
(435 to 400)

Ordovician
(500 to 435)

Cambrian
(570 to 500)

Proterozoic
(2500 to 570)

Archean
(4600 to 2500)

The Cambrian period from 570 to 500 million years ago was named for a mountain range in central Wales, Great Britain, that contained sediments with the earliest known fossils. Nineteenth-century geologists were often puzzled by ancient rocks that were practically devoid of fossils until the Cambrian period, when life seemingly sprang up in great abundance the world over. The base of the Cambrian was thought to mark the beginning of life, and all time before then was simply called Precambrian.

The period was generally quiet in terms of geologic processes, with little mountain building, volcanic activity, glaciation, or extremes in climate. The breakup of the late Precambrian supercontinent and the flooding of continents with inland seas created abundant warm shallow-water habitats, prompting an explosion of new species. Never before or since have so many novel and unusual organisms existed; surprisingly, none have any counterpart in today's living world.

THE CAMBRIAN EXPLOSION

The Cambrian was an evolutionary heyday for species, featuring a 5- to-10-million-year explosion of the first complex animals with exoskeletons. The early Cambrian witnessed the disappearance of soft-bodied Ediacaran faunas and the proliferation of shelly faunas. The biologic proliferation peaked about 530 million years ago, filling the ocean with a rich assortment of life. Seemingly out of nowhere and in bewildering abundance, animals appeared in an astonishingly short time span cloaked in a baffling variety of exoskeletons.

The introduction of hard skeletal parts has been called the greatest discontinuity in Earth history and signaled a major evolutionary change by accelerating the developmental pace of new organisms. Nearly all major groups of modern animals appeared in the fossil record, and for the first time animals sported shells, skeletons, legs, and sensing antennae. So many new and varied life forms came into existence the age is depicted as the "Cambrian explosion."

The period follows on the heels of the great Precambrian ice age, the worst the world has ever experienced, when ice sheets covered half the landmass. It was also a time when oceanic oxygen concentrations rose to significant levels. After the ice retreated and the seas began to warm, life took off in all directions. Unique and bizarre creatures roamed the ocean depths, and the Cambrian saw a higher percentage of experimental organisms than any other interval of geologic history, with perhaps three times more phyla in existence than today.

At the beginning of the Cambrian, an ocean turnover might have brought unusual amounts of nutrient-rich bottom water to the surface. Due to an increase in seawater calcium levels, early soft-bodied creatures developed skeletons as receptacles for the disposal of excess amounts of this toxic mineral from their tissues. As concentrations of calcium in the ocean further increased, animal skeletons became more diverse and elaborate.

Levels of atmospheric oxygen appeared to rise in concert with the skeletal revolution. The higher oxygen levels increased metabolic energy, enabling the growth of larger animals, which in turn required stronger structural supports. Skeletons also evolved as a response to an incoming wave of fierce predators. Paradoxically, most of these predators were soft-bodied and therefore did not survive as fossils.

Soft-bodied organisms living prior to the arrival of shelled animals at the beginning of the Cambrian had a great difficulty entering the fossil record. Animals with soft body parts decayed rapidly upon death, and so only traces of their existence remain, such as imprints, tracks, and borings. Fossil impressions of soft-bodied animals in the Ediacara Hills of southern Australia date from about 670 million years ago.

At the dawn of the Cambrian, 100 million years after the appearance of the Ediacaran fauna, most of which were evolutionary dead ends, the seascape abruptly changed with the sudden arrival of animals with hard skeletal parts. Most phyla of living organisms appeared almost simultaneously, many of which had their origins in the latter part of the Precambrian.

Body styles that evolved in the Cambrian largely served as blueprints for modern species, with few new radical body plans appearing since then.

When skeletons evolved, the number of organisms preserved in the fossil record jumped dramatically. All known animal phyla that readily fossilized appeared during the Cambrian, after which the number of new classes decreased sharply. Fossils became so abundant at the beginning of the period for several reasons: the development of hard body parts that fossilize by replacement with calcium carbonate or silica, rapid burial that prevents attack by scavengers and decay by oxidation, long periods of deposition with little erosion, and large populations of species.

THE AGE OF TRILOBITES

The Cambrian is best known for a famous group of invertebrates called trilobites (Fig. 4-1), which were primitive crustaceans and a favorite among fossil collectors. The trilobite body has three lobes (hence the name), consisting of a central axial lobe containing the creature's essential organs and two side, or pleural, lobes. Since trilobites were so widespread and lived throughout the Paleozoic, their fossils are important markers for dating rocks of this era. They appeared at the very base of the Cambrian and became the dominant invertebrates of the early Paleozoic, diversifying into about 10,000 species before declining and becoming extinct after some 340 million years of highly successful existence.

About 10 million years into the Cambrian, a wave of extinctions decimated a huge variety of newly evolved species. The mass extinction eliminated more than 80 percent of all marine animal genera and is numbered among the worst in Earth history. It coincided with a drop in sea level that followed continental collisions. The die-offs paved the way for the ascendancy of the trilobites, which were among the first organisms to grow hard shells and the dominant species for the next 100 million years. They were small, oval-shaped arthropods ancestors to the horseshoe crab, their only remaining direct descendant. Only the giant paradoxides was truly a paradox among trilobites, extending nearly 2 feet in length, while most trilobites were less than 4 inches long.

The trilobites occupied the shallow waters near the shores of ancient seas that flooded inland areas, providing extensive continental margins from the coastline to the abyss. In addition, a stable environment enabled marine life to flourish and proliferate. The flora included cyanobacteria, red and green algae, and acritarchs, a form of plankton that supported early Paleozoic food chains.

Curiously, many trilobite fossils have bite scars predominantly on the right side. Predators might have attacked from the right because when the trilobite curled up to protect itself, it exposed this side of its body. (Trilobite fossils are often found with their bodies completely curled up.) However, if the trilobite had a vital organ on its left side and an attack occurred there, it stood a good chance of being eaten, leaving no fossil. Therefore, those attacked on the right side stood a better chance of entering the fossil record. Trilobites shed their exoskeletons as they grew, and in this manner an individual could have left several incomplete fossils, which explains why whole fossils are rare.

The population of trilobites peaked during the late Cambrian around 520 million years ago, when they accounted for about two-thirds of all marine species. By the mid-Ordovician, about 460 million years ago, that fraction dropped to one-third because of the rise of mollusks and other competing animals. Later, the trilobites left the nearshore for the offshore, possibly due to environmental changes. The decline of the trilobites also might be connected with the arrival of the jawed fishes.

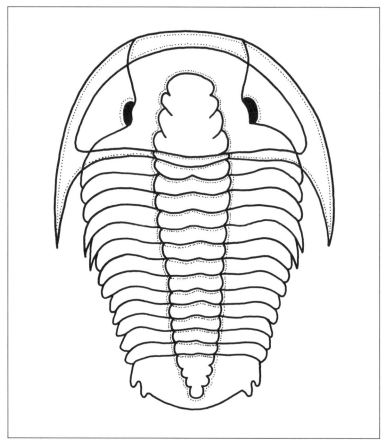

Figure 4-1   Trilobites are the extinct ancestors of today's horseshoe crabs.

CAMBRIAN PALEONTOLOGY

The coelenterates, which were well represented in the Cambrian seas, are the most primitive of animals and include jellyfish, sea anemones, sea pens, and corals. Most coelenterates are radially symmetrical, with body parts radiating outward from a central point. They have a saclike body and a

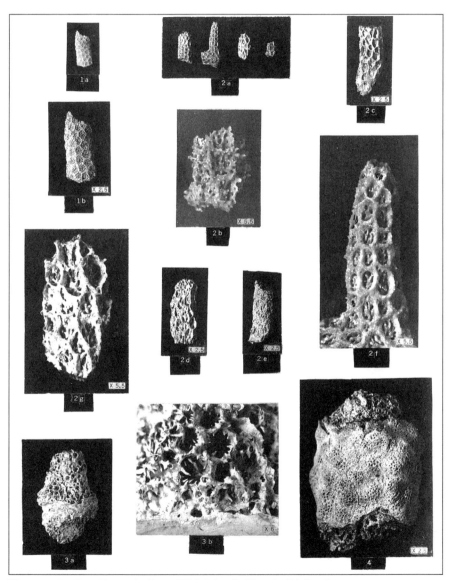

Figure 4-2   Fossil corals from Bikini Atoll, Marshall Islands. Photo by J. W. Wells, courtesy of USGS

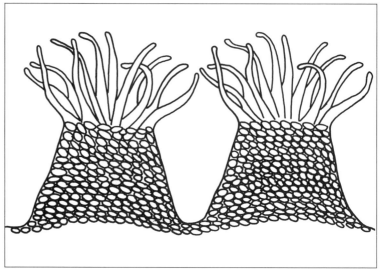

Figure 4-3   Coral polyps seek protection from predators and low tide in carbonate cups.

mouth surrounded by tentacles. The corals possess a large variety of skeletal forms (Fig. 4-2), and successive generations built thick limestone reefs. Corals began constructing reefs in the lower Paleozoic, forming chains of islands and barrier reefs along the shorelines of the continents.

The archaeocyathans resembled both corals and sponges but have no close relationship to any living group and thus belong in their own unique phylum. They formed the earliest reefs, eventually becoming extinct in the Cambrian. Many corals declined and were replaced by sponges and algae in the late Paleozoic due to the recession of the seas in which they once thrived.

The coral polyp is a soft-bodied creature that is essentially a contractible sac, crowned by a ring of tentacles (Fig. 4-3). The tentacles surround a mouth-like opening and are tipped with poisonous stingers. The polyps live in individual skeletal cups composed of calcium carbonate. They extend their tentacles to feed at night and withdraw during the day or at low tide to keep from drying in the sun. The corals coexist symbiotically (in conjunction) with zooxanthellae algae, which live within the polyp's body. The algae consume the coral's waste products and produce organic materials that nourish the polyp. Because the algae require sunlight for photosynthesis, corals must live in warm, shallow water.

The echinoderms, meaning "spiny skin," were perhaps the strangest animals ever preserved in the fossil record of the early Paleozoic. Their five-fold radial symmetry with arms radiating outward from the center of the body makes them unique among the more complex animals. They are

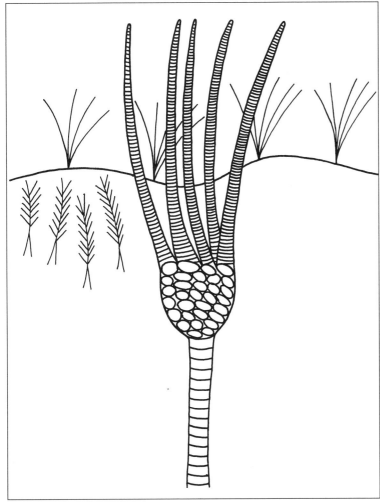

Figure 4-4   Crinoids were a dominate species in the middle and late Paleozoic and are still in existence today.

the only species possessing a water vascular system composed of internal canals that operate a series of tube feet, or podia, used for locomotion, feeding, and respiration. The echinoderms have more classes than any phylum both living and extinct. The major classes of living echinoderms include starfish, brittle stars, sea urchins, sea cucumbers, and crinoids (Fig. 4-4), known as sea lilies because of their plantlike appearance.

The brachiopods, also called lamp shells, were once the most abundant and diverse organisms, with more than 30,000 species cataloged in the fossil record. They had two saucerlike shells called valves that fitted face to face and opened and closed using simple muscles. More advanced species, including living brachiopods called articulates, had ribbed shells with interlocking teeth that opened and closed along a hinge. The brachiopods were fixed to the ocean floor by a rootlike appendage and filtered food particles through opened shells that closed to protect the animals against predators. The appearance of abundant brachiopod fossils indicates the existence of seas of moderate to shallow depth.

The mollusks are a highly diverse group and left the most impressive fossil record of all marine animals (Fig. 4-5). The phylum is so diverse it is often difficult to find common features among its members. The three major groups are the snails, clams, and cephalopods. Snails and the related slugs comprise the largest group and ranged from the Cambrian onward. The cephalopods, which include the squid, cuttlefish, octopus, and nautilus, traveled by jet propulsion. They sucked water into a cylindrical cavity through openings on each side of the head and expelled it under pressure through a funnel-like appendage. Their straight, streamline shells, up to 30 feet and more in length, made the nautiloids among the swiftest animals of the ancient seas. The ammonoids (Fig. 4-6) were the most spectacular of marine predators, with a large variety of coiled shell forms.

Figure 4-5   Molds and shells of mollusks on highly fossiliferous sandstone of the Glenns Ferry Formation on Deadman Creek, Elmore County, Idaho. Photo by H. E. Malde, courtesy of USGS

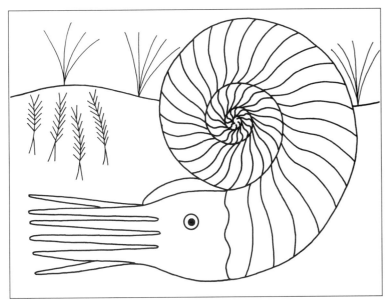

Figure 4-6 The ammonoids, ancestors of the nautilus, were aggressive predators in the ancient seas.

The arthropods are the largest phylum of living organisms, comprising roughly a million species, or about 80 percent of all known animals. Three-foot-long arthropods represent one of the largest of all Cambrian invertebrates. Among the first and best known of the ancient arthropods were the trilobites. The arthropod body is segmented, suggesting a relationship to the annelid worms. Paired, jointed limbs generally present on most segments were modified for sensing, feeding, walking, and reproduction. The crustaceans are primarily aquatic arthropods and include shrimp, lobsters, crabs, and barnacles.

The conodonts are fossilized tiny jawlike appendages, commonly occurring in marine rocks ranging from the Cambrian to the Triassic and important for dating Paleozoic rocks. They are among the most puzzling of all fossils and are thought to be parts of an unusual, soft-bodied animal, perhaps the most primitive of the vertebrates. Graptolites were colonies of cupped organisms that resembled stems and leaves, appearing much like plants but actually being animals. They either clung to the seafloor like small shrubs, floated freely near the surface, resembling tiny hacksaw blades, or attached themselves to seaweed.

Large numbers of graptolites buried in the bottom mud produced organic-rich black shales that indicate poor oxygen conditions and are important markers for correlating rock units of the lower Paleozoic. They were thought to have gone extinct in the late Carboniferous about 300 million years ago,

but the discovery of living pterobranchs, possible counterparts of graptolites, suggests these might be "living fossils."

THE BURGESS SHALE FAUNA

The Burgess Shale Formation in British Columbia, Canada contains the
remains of bizarre soft-bodied animals that appeared about 530 million
years ago, soon after the emergence of complex creatures. Some organisms
might be surviving Ediacaran fauna, most of which became extinct near the
end of the Precambrian. Indeed, the so-called Cambrian explosion might
have been triggered in part by the availability of habitats vacated when the
Ediacaran species departed. Though many mass extinctions of marine
organisms have occurred since then (Table 4-1), no fundamentally new
body plans have appeared during the past 500 million years.

The Burgess Shale invertebrates have specialized adaptations and are
surprisingly complex. However, most species became extinct later in the
Cambrian, and only a few gave rise to anything living today. Many of these
bizarre animals, some of which were possible carryovers from the upper
Precambrian, never made it beyond the middle Paleozoic. They were so

TABLE 4–1 RADIATION AND EXTINCTION OF SPECIES

Organism	Radiation	Extinction
Mammals	Paleocene	Pleistocene
Reptiles	Permian	Upper Cretaceous
Amphibians	Pennsylvania	Permian–Triassic
Insects	Upper Paleozoic	
Land plants	Devonian	Permian
Fishes	Devonian	Pennsylvanian
Crinoids	Ordovician	Upper Permian
Trilobites	Cambrian	Carboniferous and Permian
Ammonoids	Devonian	Upper Cretaceous
Nautiloids	Ordovician	Mississippian
Brachiopods	Ordovician	Devonian and Carboniferous
Graptolites	Ordovician	Silurian and Devonian
Foraminiferans	Silurian	Permian and Triassic
Marine invertebrates	Lower Paleozoic	Permian

strange, they continue to defy efforts to classify them into existing taxonomic groups.

One peculiar animal appropriately named hallucigenia (Fig. 4-7) propelled itself across the seafloor on seven pairs of pointed stilts. Seven tentacles arose from the upper body, and each appears to have had its own individual mouth. Another curious Burgess Shale animal called wiwaxia was a spiny creature about an inch long, possibly related to a modern scaleworm known as a sea mouse. An unusual worm had enormous eyes and prominent fins. Another odd creature had five eyes arranged across its head, a vertical tail fin to help steer it across the seafloor, and a grasping organ projected forward for catching prey.

An extraordinary arthropod called anomalocarids (Fig. 4-8) was possibly the largest of the Cambrian predators, reaching 3 feet in length. Its mouth was surrounded by spiked plates and flanked by a pair of jointed appendages apparently designed for holding and crushing the armored plates of invertebrates. The animal appears to have been well equipped for devouring crustaceans and was appropriately dubbed the "terror of trilobites." Several trilobite species evolved long spines that might have served as protection against anomalocarid attacks.

The Burgess Shale faunas originated in shallow water on a gigantic coral reef covered with mud that surrounded the continent of Laurentia, which included the present United States, Canada, and Greenland. Most came from the western Cordillera of North America, an ancient mountain range

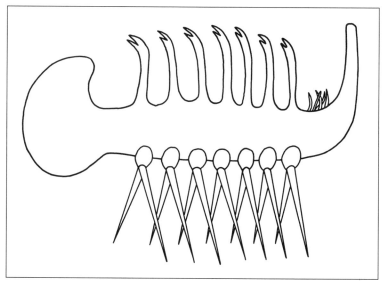

Figure 4-7 **Hallucigenia is one of the strangest animals preserved in the fossil record. It walked on 7 pairs of stilts and had 7 tentacles, each with its own head.**

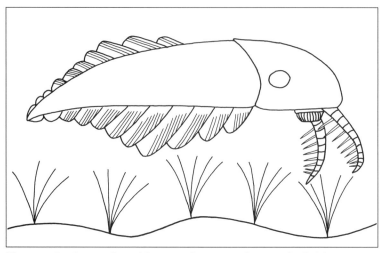

Figure 4-8   Anomalocarids were fierce predators of trilobites.

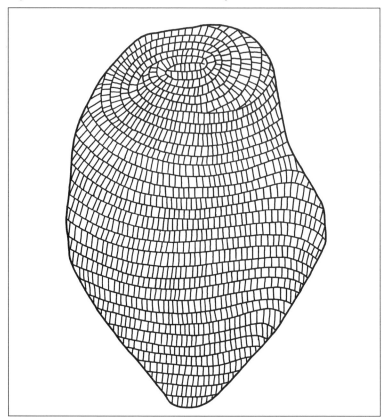

Figure 4-9   Helicoplacus was a novel species whose body parts were arranged in a manner not found in any living creature; it became extinct about 510 million years ago.

that faced open ocean in the middle Cambrian. Their widespread distribution around other continents suggests that many members could swim. Most of the Burgess Shale faunas abruptly went extinct at the end of the Cambrian, and only a few archaic forms survived to the middle Devonian.

GONDWANA

Near the end of the Precambrian, roughly 700 million years ago, all landmasses assembled into a supercontinent. The continental collisions resulted in environmental changes that had a profound influence on the evolution of life. No broad oceans or extreme differences in temperature existed to prevent species from migrating to various parts of the world. Between 630 and 560 million years ago, the supercontinent rifted apart and four or five continents rapidly drifted away. Evidence for the breakup exists in a long belt of volcanism near the present Appalachians.

Most of the continents were near the equator, which explains the existence of warm Cambrian seas. The continental breakup caused sea levels to rise and flood large portions of the land at the beginning of the Cambrian. The extended shoreline might have spurred the explosion of new species, with twice as many phyla living during the Cambrian than before or since.

Many organisms were in existence, none of which have any modern counterparts. One example is helicoplacus (Fig. 4-9), whose body parts were configured in a manner not found in any living organism. It was about

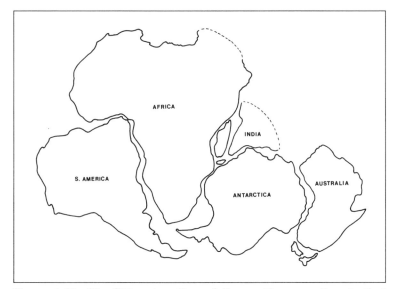

Figure 4-10 The fitting together of the southern continents that comprised Gondwana.

two inches long and shaped like a spindle covered with a spiraling system of armored plates. It emerged during the transition from the Precambrian to the Cambrian, when more types of body plans arose than at any other time. Helicoplacus, like most species of the early Cambrian, was unsuccessful in the long run and became extinct about 510 million years ago, just 20 million years after it first appeared.

During the Cambrian, continental motions assembled the present continents of South America, Africa, India, Australia, and Antarctica into Gondwana (Fig. 4-10), named for an ancient region of east-central India. Evidence for Gondwana exists in geologic provinces with similar rock types from the late Precambrian to the early Cambrian; these show matches between Brazil and west Africa; eastern South America, South Africa, west Antarctica, and east Australia; and east Africa, India, east Antarctica, and

Figure 4-11   Lystrosaur fossil sites in Gondwana.

west Australia. A great mountain-building episode deformed areas between all pre-Gondwana continents, indicating their collision during this interval.

Much of Gondwana was in the south polar region from the Cambrian to the Silurian. The present continent of Australia was at the northern edge of Gondwana and located on the Antarctic Circle. A later collision between North America and Gondwana near the end of the Cambrian about 500 million years ago created an ancestral Appalachian range that continued into western South America long before the Andes formed. North America then broke away from Gondwana and linked with Greenland and Eurasia to form Laurasia about 400 million years ago. Eurasia, the largest modern continent, assembled with about a dozen individual continental plates that welded together at the end of the Proterozoic.

A preponderance of evidence for the existence of Gondwana includes fossilized finds of a mammal-like reptile called lystrosaurus in the Trans-antarctic Range of Antarctica, which indicates a link with southern Africa and India, the only other known sources of lystrosaurus fossils (Fig. 4-11). A fossil of a South American marsupial in Antarctica, which acted as a land bridge between the southern tip of South America and Australia, lends additional support to the existence of Gondwana.

Further evidence for Gondwana includes fossils of a reptile called mesosaurus in eastern South America and South Africa. Fossils of the late Paleozoic fern glossopteris, from the Greek word meaning featherlike and whose leaf impressions actually look like feathers, exist in coal beds on the southern continents and India. However, the plant is suspiciously absent on the northern continents, suggesting the existence of two large conti-nents, one located in the Southern Hemisphere and another in the Northern Hemisphere, separated by a large open sea.

EONS	PERIODS
	(all dates: millions of years ago)
Phanerozoic (570 to present)	Quaternary (3 to present)
	Late Tertiary (25 to 3)
	Early Tertiary (65 to 25)
	Cretaceous (135 to 65)
	Jurassic (210 to 135)
	Triassic (250 to 210)
	Permian (280 to 250)
	Carboniferous (345 to 280)
	Devonian (400 to 345)
	Silurian (435 to 400)
	Ordovician (500 to 435)
	Cambrian (570 to 500)
Proterozoic (2500 to 570)	
Archean (4600 to 2500)	

5

ORDOVICIAN VERTEBRATES

The Ordovician period from 500 to 435 million years ago was named for the ancient Ordovices tribe of Wales, Great Britain. Ordovician marine sediments are recognized on all continents of the Northern Hemisphere, in the Andes Mountains of South America, and in Australia, but they are absent in Antarctica, Africa, and India. Ordovician terrestrial deposits are not easily recognized because of the lack of fossilized land organisms.

The multitudes of species that exploded onto the scene in the early Cambrian advanced significantly in the warm Ordovician seas (Fig. 5-1). Corals began building extensive carbonate reefs in the Ordovician. The first fish appeared in the ocean, and the existence of freshwater jawless fish suggests that unicellular plants, including red and green algae, were inhabiting lakes and streams on land.

THE JAWLESS FISH

Beginning about 500 million years ago, the first vertebrates appeared on the scene with an internal skeleton made of bone or cartilage, one of life's most significant advancements. The vertebrate skeleton was light, strong, and

Figure 5-1   Marine flora and fauna of the late Ordovician. Courtesy of Field Museum
of Natural History, Chicago

flexible, with efficient muscle attachments, and the skeleton grew along with the animal. These new skeletons enabled the wide dispersal of free-swimming species into a variety of environments.

Invertebrates, supported by external skeletons, were at a distinct disadvantage in terms of mobility and growth. Many animals, like crustaceans, shed their shells as they grew, which often made them vulnerable to predators. One such predator was an extinct giant sea scorpion with immense pincers called eurypterid (Fig. 5-2), which ranged from the Ordovician to the Permian and grew to 6 feet long, terrorizing shell-less creatures on the ocean floor.

The earliest vertebrates were probably wormlike creatures with a prominent rod called a notochord down the back, a system of nerves along it, and rows of muscles attached to the backbone and arranged in a banded pattern. Rigid structures made of bone or cuticle acted as levers, and with flexible joints they efficiently translated muscle contractions into organized movements such as rapid lateral flicks of the body to propel an animal through water. Later, a tail and fins evolved for stabilization, and the body became more streamlined and torpedo-shaped for speed. With intense competition among the stationary and slow-moving invertebrates, any advancement in mobility was advantageous to the vertebrates.

The oldest known vertebrates were primitive jawless fish called agnathans (Fig. 5-3), which first appeared in the early Ordovician about 470 million years ago. Remarkably well-preserved remains of these fish were discovered in the mountains of Bolivia, much of which was inundated by the sea in the Ordovician. Originally, fossil evidence was scant and fragmentary, and little was known of their appearance or about their evolutionary history. Earlier descriptions dismissed them as a headless, tailless mass of scales and plates or confused the head with the tail, giving the agnathans the dubious title "backwards fossil."

The widespread distribution of primitive fish fossils throughout the world suggest a long vertebrate record prior to the Ordovician. The first fish were small mud-grubbers and sea squirts lacking jaws and teeth. These ancient fish were probably poor swimmers and avoided deep water. The jawless fish were generally small (about the size of a minnow) and heavily armored with bony plates that protected the rounded head. The rest of the body was covered with thin scales that ended near a narrow tail. Although well protected from invertebrate predators, the added weight required the fish to live mostly on the bottom, where they sifted mud for food particles

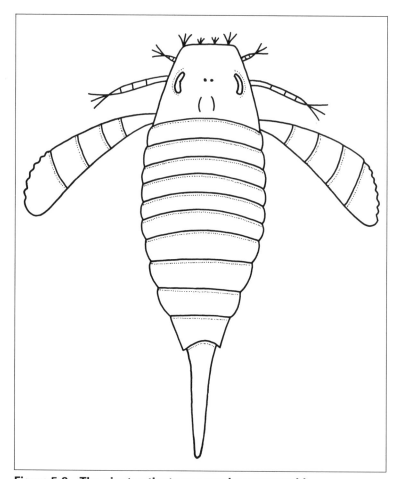

Figure 5-2   The giant extinct sea scorpion eurypterid.

Figure 5-3   The jawless fish known as agnathans were the progenitors of fish today.

and expelled the waste products through slits on both sides of the throat, which later became gills.

The jawless fish whose modern counterparts include lampreys and hagfish had a flexible rod similar to cartilage instead of a bony spine typical of most vertebrates. Gradually, the protofish acquired jaws and teeth, the bony plates gave way to scales, lateral fins developed on both sides of the lower body for stability, and air bladders maintained buoyancy. Some fish were surprisingly large, up to 18 inches long and 6 inches wide. Thus, for the first time, vertebrates skillfully propelled themselves through the sea, and the fishes soon became masters of the deep.

FAUNA AND FLORA

Corals are marine coelenterates attached to the ocean floor (Fig. 5-4). They began constructing extensive limestone reefs in the Ordovician, building chains of islands and altering the shorelines of the continents. Bryozoans, often called moss animals, are similar to corals but much smaller. They consist of microscopic individuals living in small colonies up to several inches across, giving the ocean floor a mosslike appearance. The bryozoans are retractable creatures, encased in a calcareous vaselike structure. Ciliated tentacles surround the polyp, forming a netlike structure around the mouth used for filtering microscopic food that floats by.

Fossil bryozoans are common in Paleozoic formations, especially those of the American Midwest. They resembled modern descendants of bryozoans, and some larger groups might have contributed to Paleozoic reef-building. They are most abundant in limestone and less so in shales and sandstones. Fossil bivalve shells (brachiopod and pelecypod) are often encrusted with a delicate outline of bryozoans. Because of their small size, bryozoans make ideal microfossils for dating oil well cuttings. Their abundance from the Ordovician to the present makes bryozoans highly useful for rock correlations.

Of particular importance to geologists are the ostracods, or mussel shrimp, whose fossils are useful for correlating rocks from the Ordovician onward. Starfish were also common and left fossils in the Ordovician rocks of the central and eastern United States. Their skeletons comprised tiny silicate or calcite plates that were not rigidly joined and therefore usually disintegrated when the animal died, making whole starfish fossils rare. The sea cucumbers have large tube feet modified into tentacles and skeletons comprised of isolated plates that are occasionally found as fossils.

Near the end of the Ordovician 450 million years ago, the concentration of atmospheric oxygen generated sufficient levels of ozone in the upper stratosphere to shield the Earth from the sun's deadly ultraviolet rays (Fig. 5-5). Thus, for the first time, plants began to come ashore to populate the land. When the early plants first left the oceans and lakes for a home on dry land, they were greeted by a harsh environment, where ultraviolet

Figure 5-4 A collection of corals at Saipan, Mariana Islands. Photo by P. E. Cloud, courtesy of USGS

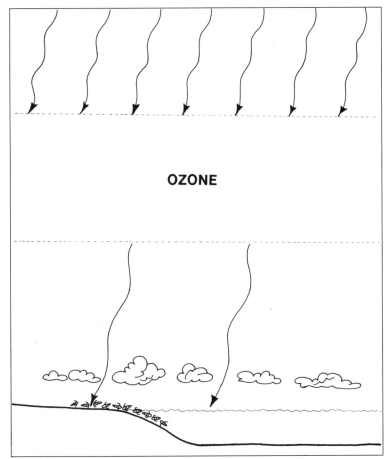

Figure 5-5 The ozone layer filters out harmful ultraviolet radiation from the sun and allows land creatures to survive.

radiation, desert conditions, and lack of nutrition made life difficult. First to greet the land plants were soil bacteria that churned sediments into lumpy brown mounds. Their presence helped speed weathering processes, without which hot bare rock would have covered most of the landscape and land plants would have had little success gaining a root hold.

Cyanobacteria, incorrectly called blue-green algae, might have been preparing the soil for the land invasion as early as 3 billion years ago. Ancient cyanobacteria, which were resistant to high levels of ultraviolet radiation, first lived in shallow tide pools, from which they eventually colonized the continents. They might have improved the terrestrial climate for a life out of water by drawing down atmospheric carbon dioxide, thereby countering the greenhouse effect, which prevents thermal energy from escaping the planet. The soil bacteria helped resist erosion by binding the

Figure 5-6 Rafts of blue-green algae lying in a pool in Indian Canyon, Duchesne County, Utah. Photo by W. H. Bradley, courtesy of USGS

sediment grains together and soaking up rainwater. Bacteria also provided nutrients for the early land plants.

Plant fossils of Ordovician age appear to be almost entirely composed of algae similar to present-day algal mats found on seashores and at the bottoms of ponds (Fig. 5-6). Some marine algae lived in the intertidal zones, and were able to withstand only short periods out of the sea, due to the risk of dehydration. Even after developing protective measures to help the organisms survive out of water for longer periods, they still depended on the sea for reproduction.

Lichens, which are a symbiotic relationship between algae and fungi, in which one lives off the other, probably took the first tentative steps on land.

Following the lichens were mosses and liverworts. Fungi also had a symbiotic relationship with the roots of plants when they first evolved, aiding vegetation with the uptake of nutrients and receiving carbohydrates in return.

Bacteria also played an important role in the fixation of nitrogen, an abundant soil gas and an important nutrient for plant life. In what is called the nitrogen cycle, denitrifying bacteria convert dissolved nitrate back into elemental nitrogen. Without this cycle, all nitrogen in the atmosphere would have long ago disappeared and the planet would have only a fraction of its present atmospheric pressure.

THE ORDOVICIAN ICE AGE

Plants began to invade the land and extend to all parts of the world during the late Ordovician about 450 million years ago. The early land plants absorbed large quantities of atmospheric carbon dioxide, and rapid burial under anaerobic conditions deposited the organic carbon into the geologic column, where it became coal. Plants also aided the weathering process, which leached minerals from the rocks, and locked up massive amounts of carbon dioxide in carbonate rocks such as limestone deposited by shelly organisms from the Cambrian onward (Fig. 5-7).

Figure 5-7   Folded Cambrian limestone on the south side of Scapegoat Mountain, Lewis and Clark County, Montana. Photo by M. R. Mudge, courtesy of USGS

The withdrawal of substantial amounts of carbon dioxide from the atmosphere weakened the greenhouse effect. The resulting climate cooling initiated in large part by the plant invasion spawned a major ice age at the end of the Ordovician about 440 million years ago. The glaciations of the late Ordovician and the glacial epochs of the middle and late Carboniferous about 330 million and 290 million years ago might have been influenced by a reduction of atmospheric carbon dioxide to about one-quarter of its present value.

Atmospheric scientists have amassed information on global geochemical cycles to ascertain the cause of such a radical change in the carbon dioxide

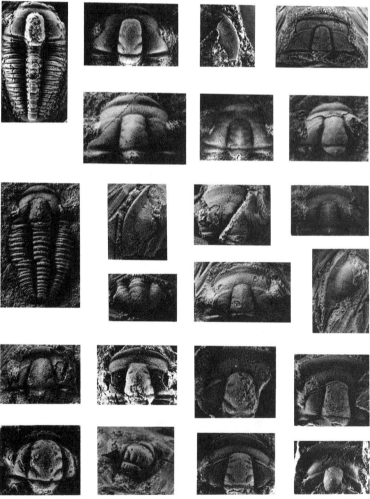

Figure 5-8   Trilobites of the Carrara Formation in the southern Great Basin, California. Photo by A. R. Palmer, courtesy of USGS

content of the atmosphere. Data from deep-sea cores show that carbon dioxide variations preceded changes in the extent of the more recent ice ages, therefore earlier glacial epochs might have been similarly affected. The variations of carbon dioxide levels might not be the sole cause of glaciation. But when combined with other processes, such as variations in the Earth's orbital motions or a drop in solar radiation, they could become a strong influence.

Continental movements also might be responsible for the late Ordovician glaciation. Magnetic orientations in rocks from many parts of the world indicate the positions of continents relative to the magnetic poles at various times in Earth history. Paleomagnetic studies in Africa revealed very curious findings, however, with the data placing North Africa directly over the South Pole during the Ordovician.

Additional evidence for such widespread glaciation came from another surprising location—the middle of the Sahara Desert. Geologists exploring for petroleum in the region stumbled upon a series of giant grooves cut into the underlying strata by roving glaciers. Rocks embedded at the base of glaciers scoured the landscape as the ice sheets moved back and forth. Other collaborating evidence that thick sheets of ice blanketed the Sahara Desert include erratic boulders (placed by moving ice) and eskers, which are sinuous sand deposits from glacial outwash streams.

As the Ordovician drew to a close, a mass extinction eliminated some 100 families of marine animals. Glaciation reached its peak, with ice sheets radiating outward from a glacial center in North Africa, which then hovered directly over the South Pole. Most of the victims were tropical species sensitive to fluctuations in the environment. Among those that went extinct were many trilobite species (Fig. 5-8). Prior to the extinction, trilobites accounted for about two-thirds of all species but only one-third thereafter. The graptolites, which were colonies of cupped organisms that resembled a conglomeration of floating stems and leaves, also became extinct at the end of the Ordovician.

THE IAPETUS SEA

During the late Precambrian and early Cambrian, a proto-Atlantic Ocean called the Iapetus opened, forming extensive Cambrian inland seas. The inundation submerged most of the ancient North American continent called Laurentia and the ancient European continent called Baltica. The Iapetus Sea was similar in size to the North Atlantic and occupied the same general location about 500 million years ago (Fig. 5-9). A continuous coastline running from Georgia to Newfoundland between 570 and 480 million years ago suggests that this ancient east coast faced a wide, deep sea.

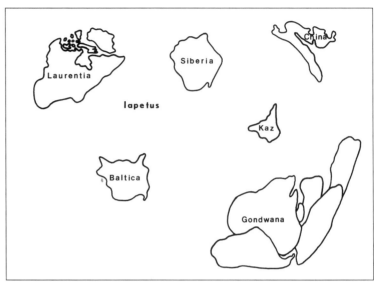

Figure 5-9 About 500 million years ago, the continents surrounded an ancient sea called the Iapetus.

The Iapetus stretched at least 1,000 miles across from east to west and bordered a much larger body of water to the south. It was dotted with volcanic islands and resembled the present-day Pacific Ocean between Southeast Asia and Australia. The shallow waters of the near-shore environment of this ancient sea from the Cambrian to the mid-Ordovician, about 460 million years ago, contained abundant invertebrates, including trilobites, which accounted for 70 percent of all species. Eventually, the trilobites faded, while mollusks and other invertebrates expanded throughout the seas.

The closing of this ancient ocean basin 400 million years ago as Baltica approached Laurentia signaled the formation of Laurasia. The island arcs between the two landmasses were scooped up and plastered against continental edges as the continents collided. The oceanic crustal plate carrying the islands subducted under, or was driven beneath, Baltica. The subduction rafted the islands into collision with the continent and deposited the formerly submerged rocks on the present west coast of Norway. Slices of land called terranes in western Europe migrated into the Iapetus from ancient Africa. In the same manner, slivers of crust from Asia have traveled across the ancient Pacific Ocean called the Panthalassa, Greek for universal sea, to form much of western North America.

A large portion of the Alaskan panhandle, known as the Alexander terrane, began its existence as part of eastern Australia some 500 million years ago. About 375 million years ago, it broke free from Australia,

Figure 5-10 Radiolarian skeletons offer clues to the histories of terranes. By studying the various shapes of these single-celled organisms, scientists can determine when and where they came from.

traversed the proto-Pacific Ocean, stopped briefly at the coast of Peru, slid past California, and rammed into the upper North American continent around 100 million years ago. The entire state of Alaska is an agglomeration of terranes that were pieces of ancient oceanic crust. Basaltic seamounts that accreted to the margin of North America traveled halfway across the ocean that preceded the Pacific.

Terranes are fault-bounded blocks, ranging in size from small crustal fragments to subcontinents, with geologic histories that are markedly different from those of neighboring blocks and of adjoining continental masses. They are a billion or more years old and are dated by analyzing fossil radiolarians (Fig. 5-10), which are marine protozoans that lived in deep water and were abundant from about 500 million to 160 million years ago. Different species also defined specific regions of the ocean where the terranes originated. Many terranes traveled great distances before finally adhering to a continental margin. Some North American terranes have a western Pacific origin and were displaced thousands of miles to the east.

North and South America apparently abutted one another at the beginning of the Ordovician (Fig. 5-11). A limestone formation in Argentina contains a distinctive trilobite species typical of North America but not of South America. The fossil evidence suggests that the two continents collided about 500 million years ago, creating an ancestral Appalachian range along eastern North America and western South America long before the present Andes formed. Later, the continents rifted apart, transferring a slice of land containing trilobite fauna from North to South America.

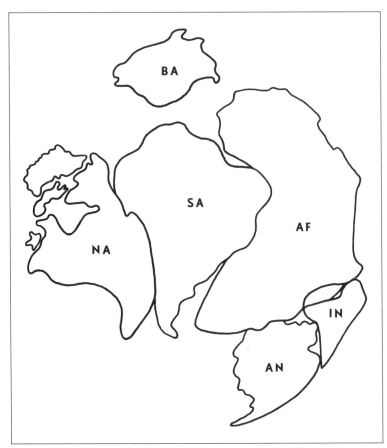

Figure 5-11 North and South America might have collided at the beginning of the Ordovician, about 500 million years ago.

THE CALEDONIAN OROGENY

The Iapetus Sea was closed off as Laurentia approached Baltica at the end of the Silurian about 400 million years ago, some 200 million years before the modern Atlantic began to open. When the continents collided, they crumpled the crust and forced up mountain ranges at the point of impact. The sutures joining the landmasses are preserved as eroded cores of ancient mountains called orogens, from the Greek word *oros* for mountain. Paleozoic continental collisions raised huge masses of rock into several folded mountain belts throughout the world (Fig. 5-12). A major mountain building episode from the Cambrian to the middle Ordovician deformed areas

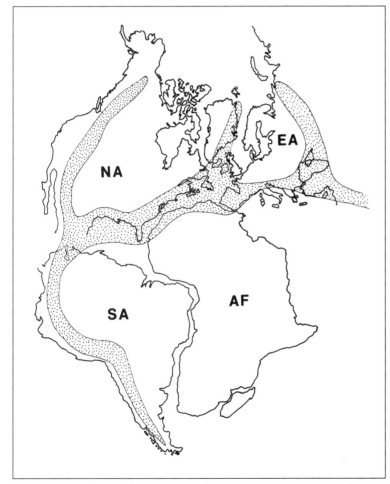

Figure 5-12   Paleozoic mountain belts developed by continental collisions.

between all continents comprising Gondwana, indicating their collision during this interval.

Matching geologic provinces exist among South America, Africa, Antarctica, Australia, and India. The Cape Mountains in South Africa have counterparts in the Sierra Mountains south of Buenos Aires in Argentina. Matches also exist between mountains in Canada, Scotland, and Norway. Much of Gondwana was in the south polar region, where glaciers expanded across the continent.

The closing of the Iapetus Sea from the middle Ordovician to the Devonian as Laurentia approached Baltica resulted in the great Caledonian orogeny, or mountain building episode. This orogenic activity formed a mountain belt that extended from southern Wales, spanned Scotland, and ran through Scandinavia and Greenland, possibly including today's extreme northwest Africa as well. In North America, this orogeny built a mountain belt that extended from Alabama through Newfoundland and reached as far west as Wisconsin and Iowa. Vermont still preserves the roots of these ancient mountains, which were shoved up between about 470 and 400 million years ago, but have since been planed off by erosion.

The middle Ordovician Taconian orogeny, named for the Taconic Range of eastern New York State, culminated in a chain of folded mountains that extended from Newfoundland through the Canadian Maritime Provinces and New England, reaching as far south as Alabama. During the Taconian disturbance, extensive volcanic activity occurred in Quebec and Newfoundland and from Alabama to New York, extending as far west as Wisconsin and Iowa.

An inland sea that flooded the continent in the middle Ordovician and reached a maximum in the late Ordovician, partially withdrew in response to a flood of sediments eroded from the Taconian mountain belts. One of these sedimentary deposits is the widespread Ordovician St. Peter Sandstone of the central United States, composed of well-sorted, nearly pure quartz beach sands, ideal for the manufacture of glass.

6

SILURIAN PLANTS

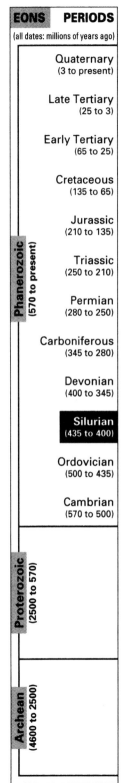

The Silurian period, which ran from 435 to 400 million years ago, was named for the Silures, an ancient Celtic tribe of Wales, Great Britain. Many of today's mountain ranges were uplifted by middle Paleozoic continental collisions. Laurentia collided with Baltica to create Laurasia and close off the Iapetus Sea during the Silurian. The collision formed mountain belts on the margins of continents surrounding the ancient sea, producing intensely folded rocks (Fig. 6-1).

Evidence of widespread reef building by Silurian corals indicates the existence of warm, shallow seas with little temperature variation. Regardless of the excellent climate, the trilobites, which were extremely successful in the early Paleozoic, began to rapidly decline in the Silurian, finally succumbing to extinction at the end of the period. The higher land plants were firmly established on the previously barren continents, and eventually creatures crawled out of the sea to dine on them.

THE AGE OF SEAWEED

The distinction between plants and animals is somewhat blurred in the distant geologic record; at one time they shared many common charac-

Figure 6-1   Folded sandstones and shales dating from the base of the upper Silurian near Hancock, Washington County, Maryland. Photo by C. D. Walcott, courtesy of USGS

teristics. From humble beginnings as simple algae during the Precambrian, single-celled plants probably colonized for the same reasons unicellular animals grouped together: structural support, division of labor, and protection. However, complex marine plants did not appear in the fossil record until the Cambrian, after which they evolved rapidly.

Although the Cambrian has been referred to as "the age of seaweed," the geologic record does not support this contention with strong fossil evidence. Well-preserved multicellular algae and a variety of fossil spores were discovered in late Precambrian and Cambrian sediments, suggesting complex sea plants had evolved, but no other significant remains have been found. Even as late as the Ordovician, plant fossils appear to have been composed almost entirely of algae, which probably formed stromatolite mounds and algal mats similar to those on seashores today.

The early seaweeds were soft and nonresistant and generally did not fossilize well. A seaweedlike plant grew half submerged in estuaries and rivers. However, for plants to be truly shore-bound, they had to reproduce entirely out of water. The first land plants achieved this function with sacs of spores attached to the ends of simple branches. When the spores matured, they were cast into the air and carried by the wind to suitable sites where they grew into new plants.

The first complex plants lived just below the surface, in shallow waters, probably as a protection against high levels of solar ultraviolet radiation. When the atmospheric oxygen content rose to near present-day levels, the upper stratospheric ozone layer began to screen out the deadly ultraviolet rays, enabling life to flourish on the Earth's surface. Once plants crept ashore (Fig. 6-2), the land was soon sprawling with lush forests.

Before the invasion of the true plants, a slimy coating of photosynthesizing cyanobacteria, or blue-green algae, might have inhabited the land. A cover of algae accelerated the weathering of rock and the formation of soils and nutrients required for advanced plant life. Therefore, prior to the emergence of land plants, microbial soils were making the Earth more hospitable for life out of water. The microorganisms probably formed a dark, knobby soil, resembling lumpy mounds of brown sugar spread over the landscape. In this manner and for about half a billion years, simple plants paved the way for their more advanced relatives.

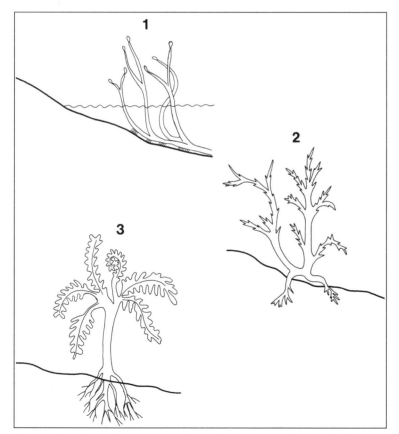

Figure 6-2   Evolution of plants from the ocean to the land, showing 1) fully aquatic, 2) semiaquatic, 3) fully terrestrial.

Prior to the arrival of the terrestrial microbes, the continents were much too hot to support complex life forms. The early organisms played an important role in cooling the land surface by drawing down the atmosphere's surplus carbon dioxide for use in photosynthesis. The loss of this potent greenhouse gas cooled the climate and allowed higher forms of life to populate the continents. The microbes also aided in the weathering of rock into soil, helped to prevent soil erosion, and provided the nutrients more advanced plants needed for survival.

The first land plants were probably algae and seaweedlike plants that lived in the intertidal zones, as well as primitive forms of lichen and moss that existed on exposed surfaces. They were followed by tiny fernlike plants, the predecessors of trees, which lacked root systems and leaves and fertilized with spores. The most complex land plants grew less than an inch tall and probably resembled an outdoor carpet covering the landscape.

By the late Silurian, all major plant phyla were in existence. Except for simple algae and bacteria, the early land plants diverged into two major groups. One gave rise to the lycopods, or club mosses, and the other gave rise to the gymnosperms, ancestors to the great majority of modern land plants. The simplest plants, among the first to live on shore, were the psilophytes, or whisk ferns, which lacked roots and leaves. They lived semisubmerged and reproduced by casting spores into the sea for dispersal.

The next major evolutionary step was the development of a vascular stem that uses channels to conduct water from a swamp or from the moist ground nearby to the plant's extremities. The strengthening of the stems enabled vascular plants to grow tall. The early club mosses, ferns, and horsetails were the first plants to make use of this system. When roots evolved plants could survive entirely on dry land by drawing water into their stalks from moist soil.

The lycopods, which included the club mosses and scale trees (Fig. 6-3), were the first plants to develop true roots and leaves. The branches were arranged in a spiral shape, and leaves were generally small. Spores were attached to modified leaves that became primitive cones. The scale trees, so named because the scars on their trunks resembled large fish scales, grew upwards of 100 feet or more high and eventually became one of the dominant trees of the Paleozoic forests.

During their first 50 million years on dry land, plants displayed increased diversity and complexity, including root systems, leaves, and reproductive organs employing seeds instead of spores. When the true leaves evolved, plants developed a variety of branching patterns to expose them to as much sunlight as possible to maximize photosynthesis. Competition for light was of primary importance to the evolution of land plants, and those that developed the most efficient branching patterns gathered the most light and were the most successful.

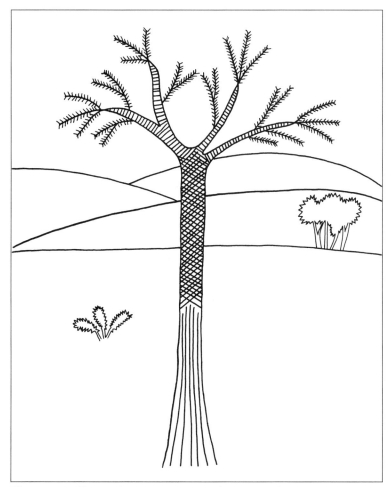

Figure 6-3   The scale tree was one of many early trees in the lush Carboniferous forest.

As plants grew larger, they progressed from random branching to tiers of branches to achieve greater efficiency with a minimum of self-shading, similar to present-day pines. This added weight increased the mechanical stress on the plant, requiring stronger branches to prevent limbs from breaking during storms. These innovations gave rise to most flora in existence today.

THE REEF BUILDERS

Silurian marine invertebrates (Fig. 6-4) were intermediate in evolution between Ordovician and Devonian lines. Coral reef formation during the

Figure 6-4 Marine flora and fauna of the middle Silurian. Courtesy of Field Museum of Natural History, Chicago

Silurian was widespread, indicating the presence of warm shallow seas with little seasonal temperature variation. Corals began constructing reefs in the Ordovician, forming barrier islands and island chains. They also built atolls atop extinct submerged volcanoes. As the volcanoes subsided beneath the sea, the rate of coral growth matched the rate of ocean subsidence, which kept the corals at a constant shallow depth for photosynthesis.

The coral polyp is a soft-bodied animal crowned by tentacles surrounding a mouthlike opening and tipped with poisonous stingers to attack prey swimming nearby. The polyp lives in an individual skeletal cup or tube composed of calcium carbonate called a theca. Polyps extend their tentacles to feed at night and withdraw into their theca during the day or at low tide to keep from drying out in the sun. Because the algae within the polyp's body require sunlight for photosynthesis, corals are restricted to warm, shallow water generally less than 100 feet deep.

A large variety of corals are well represented in the fossil record and resemble many of their modern counterparts. The tabulate corals, which became extinct at the end of the Paleozoic, consisted of closely packed polygonal or rounded thecae, some with pores covering the walls of the theca. The rugose, or horn corals (Fig. 6-5), named for their shape, were particularly abundant in the Silurian and became the major reef builders of the late Paleozoic, finally becoming extinct in the early Triassic. The hexacorals, with thecae separated by six sepia, or walls, range from the

Triassic to the present and were the major reef builders of the Mesozoic and Cenozoic eras.

In the geologic past, massive coral structures turned into some of the greatest limestone deposits on Earth. Corals constructed barrier reefs and atolls and played a major role in changing the face of the planet. Coral reefs contain abundant organic material. Many ancient reefs are composed largely of a carbonate mud with the skeletal remains of a variety of other species literally "floating" in the fine sediment, producing some of the finest fossil specimens.

Tropical plant and animal communities thrived on the reefs, due to the coral's ability to build massive, wave resistant structures. Unfortunately, these were the same species that suffered several episodes of extinction due to their narrow range of living conditions. The extinctions hit hardest those

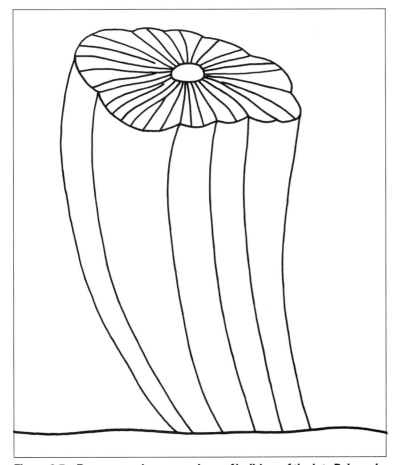

Figure 6-5   Rugose corals were major reef builders of the late Paleozoic.

organisms anchored to the ocean floor or unable to migrate out of the region because of physical and biological barriers.

The echinoderms, with their five-fold and bilateral symmetries and exoskeletons composed of numerous calcite plates, were among the most prolific animals of the Silurian seas. The most successful of the Silurian echinoderms were the crinoids, commonly called "sea lilies" because they resemble flowers atop stalks anchored to the seabed. Some crinoids were also free-swimming types. They became the dominant echinoderms of the middle and upper Paleozoic, with many species still in existence.

The long stalks of the crinoids grew upwards of 10 or more feet in length. They were composed of perhaps a hundred or more calcite disks and were anchored to the ocean floor by a rootlike appendage. A cup called a calyx perched on top of the stalk and housed the digestive and reproductive organs. The animal strained food particles from passing water currents with five feathery arms that extended from the calyx, giving the crinoid a flowerlike appearance. The extinct Paleozoic crinoids and their blastoid relatives, whose calyx resembled a rose bud, made excellent fossils, especially the stalks, which on weathered limestone outcrops often look like long strings of beads (Fig. 6-6).

The echinoids are a class of echinoderms that include sea urchins, heart urchins, and sand dollars, having exoskeletons composed of limey plates

Figure 6-6 Cogwheel-shaped crinoid columnals in a limestone bed of the Drowning Creek Formation, Fleming County, Kentucky. Photo by R. C. McDowell, courtesy of USGS

that are characteristically spiny, spherical, or radially symmetrical about a central point. Some more advanced forms were elongated and bilaterally symmetrical along a single axis. The sea urchins lived mostly among rocks encrusted with algae, upon which they fed. Unfortunately, such an environment was not conducive to fossilization. The familiar sand dollars that occasionally wash up on beaches are also rare in the fossil record.

THE LAND INVASION

The first invertebrates to crawl out of the sea and populate the land were probably crustaceans. These ancient arthropods emerged from the ocean soon after plants began to colonize the continents. The oldest known land-adapted animals were centipedes and tiny spiderlike arachnids about the size of a flea, both found in Silurian rocks some 415 million years old. The arachnids are air-breathing crustaceans and include spiders, scorpions, daddy longlegs, ticks, and mites. The early terrestrial communities probably consisted of small plant-eating arthropods that served as prey for the arachnids, which were predatory animals.

The early crustaceans were segmented creatures, ancestors of today's millipedes, and walked on perhaps 100 pairs of legs. At first, they stayed near shore, eventually moving farther inland along with the mosses and lichens. Having the land to themselves with no competitors and an abundant food supply, some species evolved into the first terrestrial giants. The crustaceans were easy prey for the descendants of the eurypterids, giant sea scorpions, when they came ashore.

The advent of forests, where leaves and other edible parts grew beyond easy reach from the ground, posed new problems for the ancestors of the insects. Climbing up tall tree trunks to feed on stems and leaves was probably less dangerous than the treacherous journey back down. It would have been much easier simply to jump or glide through the air on primitive winglike structures. These appendages probably originated as a means to regulate the insect's body temperature and by natural selection developed into flapping wings. They worked well for launching insects to the tree tops and escaping predators when the vertebrates came to shore.

Insects are by far the largest living group of arthropods. They have three pairs of legs and typically two pairs of wings on the thorax, or midsection. In most cases, the insect body is covered with an exoskeleton made of chitin, which is similar to cellulose. To achieve flight, insects had to be lightweight. As a result, their delicate bodies did not fossilize well, unless trapped in tree sap, which became hard amber, allowing the bodies to withstand the rigors of time. In some groups, the exoskeleton was composed of calcite or calcium phosphate, which enhanced the insect's chances of fossilization.

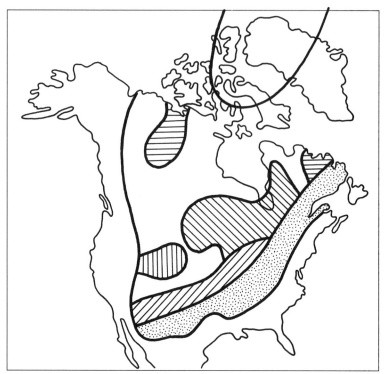

Figure 6-7 The microcontinents that came together, forming Laurentia.

LAURASIA

During the Silurian, all northern continents collided to form Laurasia. The ancestral North American continent called Laurentia was assembled from several microcontinents that collided beginning about 1.8 billion years ago (Fig. 6-7). Most of the continent evolved in a relatively brief period of 150 million years and comprised the interior of North America, Greenland, and Northern Europe. Laurentia continued to grow by garnering bits and pieces of continents and chains of young volcanic islands. After the rapid continent building, the interior of Laurentia erupted with igneous activity that lasted from 1.6 to 1.3 billion years ago.

A broad belt of red granites and rhyolites, which are igneous rocks formed by molten magma solidifying below ground and on the surface (Table 6-1), extended several thousand miles across the interior of the continent from southern California to Labrador. The Laurentian granites and rhyolites are unique due to their shear volume, which suggests that the continent stretched and thinned almost to the point of rupture, bringing mantle material near the surface. These rocks are presently exposed in Missouri,

TABLE 6–1 THE CLASSIFICATION OF VOLCANIC ROCKS

Property	Basalt	Andesite	Rhyolite
Silica content	Lowest about 50%, a basic rock	Intermediate about 60%	Highest more than 65%, an acid rock
Dark mineral content	Highest	Intermediate	Lowest
Typical minerals	Feldspar Pyroxene Olivine Oxides	Feldspar Amphibole Pyroxene Mica	Feldspar Quartz Mica Amphibole
Density	Highest	Intermediate	Lowest
Melting point	Highest	Intermediate	Lowest
Molten rock viscosity at the surface	Lowest	Intermediate	Highest
Formation of lavas	Highest	Intermediate	Lowest
Formation of pyroclastics	Lowest	Intermediate	Highest

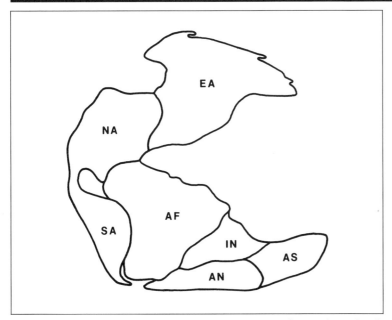

Figure 6-8   The supercontinent Pangaea was one of many large land-masses in Earth history and formed about 250 million years ago.

Oklahoma, and a few other localities, but in most of the center of the continent they are buried under sediments up to a mile thick.

These massive outpourings of igneous rocks in the interior of the continent suggest that Laurentia was part of a supercontinent that formed about 1.6 billion years ago and broke up around 1.3 billion years ago, coinciding with the igneous activity. The supercontinent acted like an insulating blanket over the upper mantle, allowing heat to collect underneath it. About 1.1 billion years ago, vast quantities of molten basalt spewed to the surface out of a huge tear in the crust running from southeast Nebraska to the Lake Superior region.

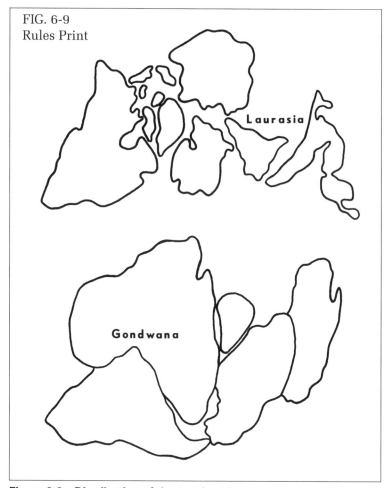

FIG. 6-9
Rules Print

Laurasia

Gondwana

Figure 6-9   Distribution of the continents 400 million years ago with the formative Laurasia in the Northern Hemisphere and Gondwana in the Southern Hemisphere.

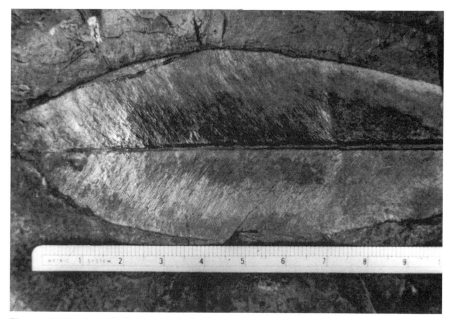

Figure 6-10   Fossil glossopteris leaf, whose existence on the southern continents is strong evidence for Gondwana. Photo by D. L. Schmidt, courtesy of USGS

About 700 million years ago, Laurentia collided with another large continent on its southern and eastern borders, creating a new supercontinent centered over the equator. A superocean located approximately in the location of the present Pacific Ocean surrounded the supercontinent. The collision thrust up a 3,000-mile-long mountain belt in eastern North America during an episode called the Grenville orogeny. A similar mountain belt occupied parts of western Europe as well.

The supercontinent rifted apart between about 630 and 560 million years ago, and its constituent continental blocks drifted away from one another. As the continents dispersed and subsided, seawater flooded the interiors, creating large continental shelves where vast arrays of organisms evolved. The rapid evolution of species at this time was highly remarkable. Another exceptional episode of explosive evolution occurred when a supercontinent called Pangaea (Fig. 6-8) rifted apart some 400 million years later.

About 500 million years ago, the continents dispersed around the Iapetus Sea, which opened during the breakup of the late Precambrian supercontinent. When the continents reached their maximum dispersal roughly 480 million years ago, subduction of the ocean floor beneath the North American plate initiated a period of volcanic activity and mountain building. From about 420 million to 380 million years ago, Laurentia collided with Baltica, the ancient European landmass, and closed off the Iapetus. The collision fused the two continents into the megacontinent of Laurasia

(Fig. 6-9), named for the Laurentian province of Canada and the Eurasian continent.

When Laurentia united with Baltica, island arcs in a proto-Pacific Ocean called the Panthalassa began to collide with the western margin of what is now North America. The collisions led to the Antler orogeny, which intensely deformed rocks in the Great Basin region, from the California-Nevada border to Idaho.

Laurasia occupied the Northern Hemisphere, while its counterpart Gondwana, which included the present Africa, South America, Australia, Antarctica, and India, inhabited the Southern Hemisphere. A large body of water called the Tethys Sea, named for the mother of the seas in Greek mythology, separated the two megacontinents. Evidence for the existence of a wide seaway between the landmasses comes from a unique specimen of flora called glossopterus (Fig. 6-10) found in the southern lands but absent in Laurasia.

The continents were lowered by erosion and shallow seas flowed inland, flooding more than half the land surface. The inland seas and wide continental margins, along with a stable environment, provided excellent growing conditions and enabled marine life to flourish and spread throughout the world.

7

DEVONIAN FISHES

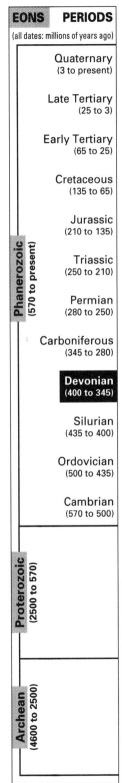

The Devonian period from 400 to 345 million years ago was named for the marine rocks of Devon in southwest England. Rocks of Devonian age exist on all continents and show widespread marine and terrestrial conditions. The period coincides with the megacontinents Laurasia and Gondwana approaching each other and pinching off the Tethys Sea between them. The wide distribution of deserts, evaporite deposits, coral reefs, and coal deposits as far north as the Canadian arctic indicates a warm climate over large parts of the world.

The warm Devonian seas spurred the evolutionary development of marine species (Fig. 7-1), including the first appearance of the ammonoids, which were coiled-shelled cephalopods that became fantastically successful in the succeeding Mesozoic era. The vertebrates, the highest form of marine animal life, dominating all other species, left their homes in the sea to establish a permanent residence on land, which by then was fully forested. Toward the end of the Devonian, the climate cooled, possibly causing glaciation near the poles. The climate change brought down many tropical marine animals, paving the way for entirely new species especially adapted to the cold.

Figure 7-1 Marine flora and fauna of the middle Devonian. Courtesy of Field Museum of Natural History, Chicago

THE AGE OF FISHES

The Devonian has been popularly named the "age of fishes." The fossil record reveals so many and varied kinds of fish that paleontologists have a difficult time classifying them all. Every major class of fish alive today had ancestors in the Devonian, but not all Devonian fishes made it to the present.

The rise of fishes in the Devonian seas contributed to the decline of their less mobile invertebrate competitors, culminating in an extinction that eliminated many tropical marine groups at the end of the period. When a mass extinction occurs, those individuals that evolve into a better adaptive form are selected for survival, which is why certain species survive one major extinction after another. This is particularly true for marine species like the sharks, which originated in the Devonian around 400 million years ago and have survived every mass extinction since.

Fish comprise over half the species of vertebrates, both living and extinct. They include the jawless fish (lampreys and hagfish), the cartilaginous fish (sharks, skates, rays, and ratfish), and the bony fish (salmon, swordfish, pickerel, and bass). The ray-finned fishes are by far the largest group of living fish species. Fish progressed from rough scales, asymmetrical tails,

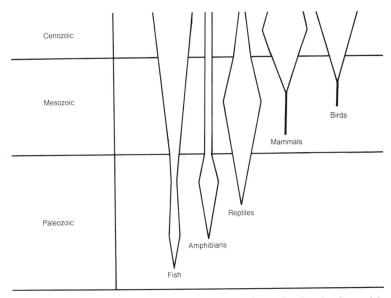

Figure 7-2 **The appearance and radiation of species beginning with the first fish in the Ordovician.**

and cartilage in their skeletons to flexible scales, powerful advanced fins and tails, and all-bone skeletons, much like they are today.

The jawless fish, which first appeared in the Ordovician (Fig. 7-2), are the earliest known vertebrates and have been in existence for 470 million years. Instead of a bony spine like most vertebrates, the jawless fish had a flexible rod similar to cartilage. They were probably poor swimmers and avoided deep water. Bony plates surrounded the head for protection from

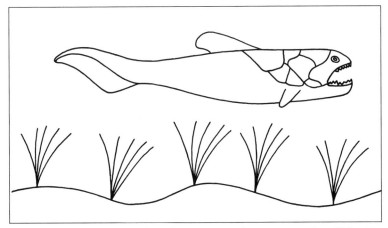

Figure 7-3 **The extinct placoderms were giants measuring 30 feet in length, and lived from early to late Devonian.**

predation, and thin scales covered the rest of the body. Although well protected from invertebrate predators, the added weight kept the fish mostly on the seafloor, where they sifted bottom sediments for food particles.

The extinct placoderms (Fig. 7-3), which reached 30 or more feet in length, were ferocious giants that preyed on smaller fishes. They had well-developed articulated jaws and thick armor plating around the head that extended over and behind the jaws. The coelacanths (Fig. 7-4) were thought to have gone extinct along with the dinosaurs 65 million years ago. However, in 1938, fishermen caught a five-foot coelacanth in the deep, cold waters of the Indian Ocean off the Comoro Islands near Madagascar.

The fish looked ancient, a castaway from the distant past, with a fleshy tail, a large set of forward fins behind the gills, powerful, square, toothy jaws, and heavily armored scales. The most remarkable aspect about this fish was that it had not changed significantly from its primitive ancestors, which evolved in the Devonian seas some 400 million years earlier, giving the coelacanth the title of "living fossil."

The coelacanth came from the same evolutionary branch that led to the land-dwelling vertebrates. Stout fins enabled the fish to crawl along on the deep ocean floor and were coordinated in a manner common in four-legged terrestrial animals. The fins moved like the legs of a lizard, with the forward appendage on each side advancing in concert with the rear appendage on the opposite side. Such an adaptation would have eased the transition from sea to land.

Figure 7-4   Modern coelacanths have not changed significantly from their ancient ancestors.

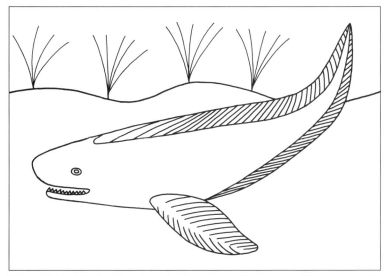

Figure 7-5   The freshwater shark xenacanthus swam like an aquatic snake.

The sharks were highly successful from the Devonian to the present. An ancient freshwater shark called Xenacanthus (Fig. 7-5) had a back fin that stretched from head to tail, and it slithered through the water like an aquatic snake. Closely related to the sharks are the rays, with flattened bodies,

Figure 7-6   Excavation for shark teeth at Shark's Tooth Hill, Kern County, California.   Photo by R. W. Pack, courtesy of USGS

pectoral fins enlarged into wings up to 20 feet across, and a tail reduced to a thin, whiplike appendage. The rays literally fly through the sea as they scoop up plankton into their mouths.

Sharks breathe by drawing water in through the mouth, passing it over the gills, and expelling it through distinctive slits behind the head. The body of the shark is heavier than water, requiring it to swim constantly or else sink to the bottom. Instead of skeletons composed of bone like most fish, shark skeletons are comprised of cartilage, a much more elastic and lighter material. However, it does not fossilize well, and about the only common remains of ancient sharks are teeth, found in marine rocks of Devonian age or younger (Fig. 7-6).

MARINE INVERTEBRATES

The Devonian marine invertebrates were similar to those that evolved in the Ordovician and include prolific brachiopods, corals, crinoids, trilo-

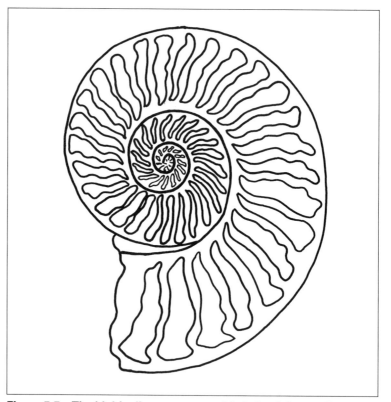

Figure 7-7 **The highly diverse ammonoids helped date sedimentary strata.**

bites, and gastropods. The conodonts, which are bony appendages of a possible leechlike animal that had the appearance of jawbones, show their greatest diversity during the Devonian and are important for long-range rock correlations of this period. Advanced insects, spiders, and poisonous centipedes appeared in the Devonian. The mollusks were well represented, and the first appearance of freshwater clams suggests that aquatic inverte- brates had already conquered the land since they only lived in estuaries and rivers.

The nautiloids and ammonoids first appeared in the early Devonian about 395 million years ago. They had external shells subdivided into air chambers, and the suture lines joining the segments presented a variety of patterns (Fig. 7-7) used for identifying various species. The air chambers provided buoyancy to counterbalance the weight of the growing shell. Most shells were coiled in a plane, some forms were coiled in a spiral, and others were essentially straight.

The now-extinct nautiloids grew upwards of 30 or more feet long, and with straight, streamlined shells they were among the swiftest and most spectacular creatures of the Devonian seas. Their neutral buoyancy (they neither sank to the bottom nor rose to the top) and ability to use jet propulsion for high-speed travel (by expelling water under pressure through a funnel-like appendage) contributed to the nautiloid's great suc- cess in the period from the Devonian to the Cretaceous.

The belemnoids, which probably originated from more primitive nau- tiloids, were abundant during the Jurassic and Cretaceous but became extinct by the Tertiary. They were related to the modern squid and octopus and possessed a long, bulletlike shell. The shell was straight in most species and loosely coiled in others. The chambered part of the shell was smaller than the ammonoid, and the outer walls thickened into a fat cigar shape.

The ammonoids were the most significant cephalopods, with a large variety of coiled shell forms (Fig. 7-8), which makes them ideal for dating Paleozoic and Mesozoic rocks. Shell designs steadily improved, making ammonoids the swiftest creatures of the deep, successfully competing with the fishes for food and avoiding predators. They lived mainly at middle depths and might have shared many features with living squids and cuttlefish. Some ammonoids grew to tremendous size, with shells up to 7 feet wide. The nautilus, which is commonly referred to as a "living fossil" because it is the only living relative of the ammonoids, lives in the depths of the South Pacific and Indian oceans down to 2,000 feet.

During a major extinction event near the end of the Devonian about 365 million years ago, many tropical marine groups disappeared, possibly due to climatic cooling. The die out of species apparently occurred over a period of 7 million years and eliminated species of corals and many other bottom- dwelling marine organisms. Primitive corals and sponges, which were

Figure 7-8 A collection of ammonoid fossils. Photo by M. Gordon Jr., courtesy of USGS

prolific limestone reef builders early in the period, suffered heavily during the extinction and never fully recovered.

While these animal groups vanished, the glass sponges, which tolerated cold conditions, diversified, only to dwindle when the crisis subsided and other groups recovered. Their prosperity during the late Devonian signifies that less fortunate species had succumbed to the effects of climatic cooling. Large numbers of brachiopod families also died out at the end of the period. In contrast, cold-adapted animals living in Arctic waters fared quite well. Much of Gondwana was in the Antarctic during the Devonian, and seas flooded broad areas of the continent. The Gondwanan fauna, which lacked reef builders and other warm-water species, survived the extinction with few losses.

The oldest species living in the world's oceans today thrive in cold waters. Many Arctic species, including certain brachiopods, starfishes, and bivalves, belong to biological orders whose origins extend hundreds of millions of years backward into the Paleozoic. In contrast, tropical faunas such as reef communities, battered by periodic mass extinctions, have come and gone quite rapidly on the geologic time scale. But not all animals that shared the same environments were identically affected by the extinction.

A possible cause for the end-Devonian extinction was the bombardment of the Earth by one or two large asteroids or comets. The meteorite impact theory is supported by the discovery of deposits containing glassy beads called microtektites in the Hunan province of China and in Belgium. Microtektites form when a large meteorite impacts the Earth and hurls droplets of molten rock into the air that quickly cool into bits of glass. The deposits also include an unusually high iridium content, which strongly indicates an extraterrestrial source. The Siljan crater in Sweden, about the same age as the microtektites, might be the source of the impact deposits. The evidence supports the notion that meteorite bombardments have contributed to many mass extinctions throughout Earth history.

TERRESTRIAL VERTEBRATES

Plants had been greening the Earth for nearly 100 million years before the vertebrates finally set foot on dry land. Previously, freshwater invertebrates and fishes inhabited lakes and streams. Freshwater fish living in Australia around 370 million years ago were almost identical to those living in China, suggesting that the two landmasses were close enough for the fish to travel between them.

By the middle Devonian, stiff competition in the sea encouraged crossopterygians to make short forays on shore to prey on abundant crustaceans and insects. The crossopterygians (Fig. 7-9) were lobe-finned fish with heavy, enamel-like scales. Their fin bones were attached to the skeleton and arranged into primitive elements of a walking limb. The crossopterygians strengthened their lobe fins into legs by digging in the sand for food and shelter. They eventually ventured farther inland, though never too distant from accessible sources of water such as swamps or streams. Primitive Devonian fish, similar to today's lungfish, crawled on their bellies from one pool to another, pushing themselves along with their fins.

Modern lungfish live in African swamps that seasonally dry out, forcing the fish to hole up for long stretches until the rains return. They burrow into the moist sand, leaving an air hole to the surface, and live in suspended animation, breathing with primitive lungs. In this manner, they can survive out of water for several months or even a year or more if necessary. When

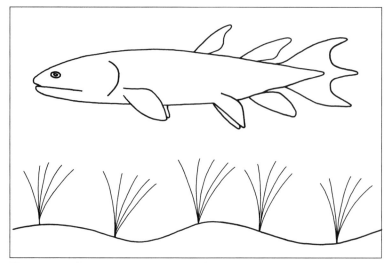

Figure 7-9   The crossopterygians, first appearing in the middle Devonian, were directly in line with air-breathing amphibians.

the rainy season returns, the pond fills again, and the fish come back to life, breathing normally with their gills.

In Florida, a walking catfish originating from Asia will leave its drying pond and travel by pushing itself along with its tail and fins, sometimes a considerable distance before finding another suitable home. They breathe with primitive nostrils and lungs as well as gills, placing them midway on the line of evolution from fish to land-living vertebrates. Air-breathing was also important for fish striving to survive in warm, shallow waters low on oxygen.

The descendants of the lobe-finned fish and lungfish were the first advanced animals to populate the land some 370 million years ago. By the late Devonian, the crossopterygians had evolved into the earliest amphibians. Their legacy is well documented in the fossil record, and at no other time in geologic history were so many varied and unusual creatures inhabiting the surface of the Earth.

One of the earliest known amphibians, named icthyostega, crawled around on primitive legs with seven toes on each hind foot. Another ancient amphibian, called acanthostega, had eight fingers on each forelimb, perhaps the most primitive of walking limbs. Amphibians sporting six and eight digits also existed, suggesting that the evolution of early land vertebrates followed a flexible pattern of development. These early amphibians living in the late Devonian, when vertebrates were first making a transition from water to land, spent most of their time in the water, which led to their eventual downfall when the great swamps dried up toward the end of the Paleozoic.

Animal tracks tell of the earliest land invasion. Tracks of the primitive Devonian fish that first ventured on dry land and gave rise to four-legged amphibians exist in formations of late Devonian age onward. The amphibian tracks are generally broad with a short stride, indicating the animal could barely hold its squat body off the ground. It walked with a clumsy gait, and running was completely out of the question.

Amphibian footprints become abundant in the Carboniferous beginning about 350 million years ago and to a lesser extent in the Permian, owing to the amphibian's preference for a life in water, in addition to the rise of the reptiles. The fossil remains of the amphibians are largely fragmentary, because of the manner by which vertebrate skeletons are constructed, with a large number of bones that are easily scattered by surface erosion, leaving a scant record of their existence.

THE OLD RED SANDSTONE

Beginning in the late Silurian and continuing into the Devonian, from about 400 million to 350 million years ago, a collision between present eastern North America and northwestern Europe raised the Acadian Mountains. The terrestrial redbeds, composed of sandstones and shales cemented by red iron oxide, of the Catskills in the Appalachian Mountains extending from southwestern New York State to Virginia are the main evidence of the Acadian orogeny in North America. Extensive igneous activity and metamorphism accompanied the mountain building at its climax. The Devonian Antler orogeny was another mountain-building episode, producing intensely deformed rocks in the Great Basin of Nevada. The Innuitian orogeny, which deformed the northern margin of present North America, resulted from a collision between the continent and another crustal plate.

The middle Devonian Old Red Sandstone, a thick sequence of chiefly nonmarine sediments in Great Britain and northwest Europe, is the main expression of this mountain building episode in Europe called the Caledonian orogeny. The formation comprises great masses of sand and mud that accumulated in the basins between the ranges of the Caledonian Mountains from Great Britain to Scandinavia. The sediments are poorly sorted and consist of red, green, and gray sandstones and gray shales that often contain fish fossils.

Erosion leveled the continents and shallow seas flowed inland, flooding more than half the landmass. The inland seas and wide continental margins, along with a stable environment, provided favorable conditions for marine life to flourish and proliferate throughout the world. Seas flooding North America during the Devonian produced abundant coral reefs that lithified (became rock) into widespread limestones.

The rising Acadian Mountains on the east side of the inland sea eroded, producing flat-lying, fossiliferous deposits of shale in western New York State, possibly the best Devonian section in the world. The vast Chattanooga Shale Formation, which covers virtually the whole continental interior, was laid down during the Devonian and Carboniferous. The seas also blanketed much of Eurasia in the late Devonian. Terrestrial clastics comprised of rock fragments eroded from the Caledonian Mountains overlay the western part of the continent.

Gondwana, to this point located in the Antarctic, now shifted its position. Its location can be shown by paleomagnetic data, which indicate the locations of continents relative to the magnetic poles by analyzing the magnetic orientations of ancient iron-rich lavas. The south magnetic pole drifted from present South Africa in the Devonian, ran across Antarctica in the Carboniferous, and ended up in southern Australia in the Permian.

The location of the southern pole is also indicated by widespread glacial deposits and erosional features on the continents that comprised Gondwana during the late Paleozoic. The mass extinctions of the middle Devonian 365 million years ago and the late Ordovician 440 million years ago coincided with glacial periods that followed long intervals of ice-free conditions.

Gondwana in the Southern Hemisphere and Laurasia in the Northern Hemisphere were separated by the Tethys Sea (Fig. 7-10), and into this seaway flowed thick deposits of sediments washed off the surrounding

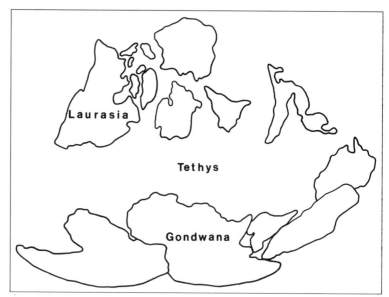

Figure 7-10   Around 400 million years ago, all continents surrounded an ancient sea called the Tethys.

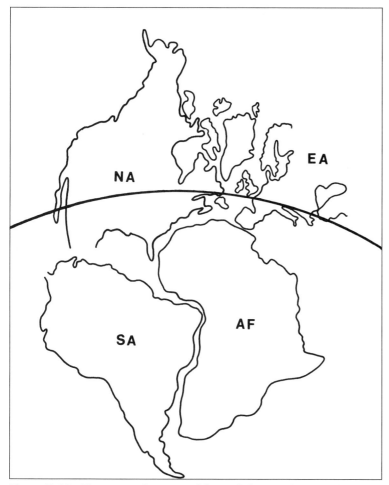

Figure 7-11 **The approximate positions of the continents relative to the equator during the Devonian and Carboniferous periods.**

continents. Their accumulated weight formed a long, deep depression in the ocean crust, called a geosyncline, which later uplifted into folded mountain belts.

A warm climate and desert conditions over large areas are indicated by widespread distribution of evaporite deposits in the Northern Hemisphere, coal deposits in the Canadian arctic, and carbonate reefs. Warm temperatures of the past are generally recognized by abundant marine limestones, dolostones, and calcareous shales. A coal belt, extending from northeastern Alaska across the Canadian archipelago to northernmost Russia, suggests that vast swamps were prevalent in these regions.

Evaporite deposits generally form under arid conditions between 30 degrees north and south of the equator. However, extensive evaporite deposits are not currently being formed, suggesting a comparatively cooler global climate. The existence of ancient evaporite deposits as far north as the arctic regions implies that either these areas were once closer to the equator or the global climate was considerably warmer in the geologic past.

The Devonian year was 400 days long and the lunar cycle was about 30.5 days as determined by the daily growth rings of fossil corals. Paleomagnetic studies indicate that the equator passed from California to Labrador and from Scotland to the Black Sea during the Devonian and Carboniferous (Fig. 7-11). The ideal climate setting helped spur the rise of the amphibians that inhabited the great Carboniferous swamps.

8

CARBONIFEROUS AMPHIBIANS

Phanerozoic (570 to present)	Quaternary (3 to present)
	Late Tertiary (25 to 3)
	Early Tertiary (65 to 25)
	Cretaceous (135 to 65)
	Jurassic (210 to 135)
	Triassic (250 to 210)
	Permian (280 to 250)
	Carboniferous (345 to 280)
	Devonian (400 to 345)
	Silurian (435 to 400)
	Ordovician (500 to 435)
	Cambrian (570 to 500)
Proterozoic (2500 to 570)	
Archean (4600 to 2500)	

The Carboniferous period, which ran from 345 to 280 million years ago, is further divided into the Mississippian and Pennsylvanian periods in North America, and was named for the coal-bearing rocks of Wales, Great Britain. Flora that appeared in the Devonian was plentiful and varied during the Carboniferous. Great coal forests of seed ferns and true trees with seeds and woody trunks spread across Gondwana and Laurasia in the lower Carboniferous.

All forms of fauna that existed in the lower Paleozoic flourished in the Carboniferous except the brachiopods, which declined in number and type. The fusulinids (Fig. 8-1), which appeared for the first time in the Carboniferous, were large, complex protozoans that resembled grains of wheat, ranging from microscopic size up to 3 inches in length. Primitive amphibians inhabited the swampy forests, which were abuzz with hundreds of different types of insects, including large cockroaches and giant dragonflies. When the climate grew colder and widespread glaciation enveloped the southern continents at the end of the period, the first reptiles emerged and displaced the amphibians as the dominant land vertebrates.

THE AGE OF AMPHIBIANS

The ancestors of the amphibians appear to have been the crossopterygians, the stem group from which all terrestrial vertebrates descended. They were

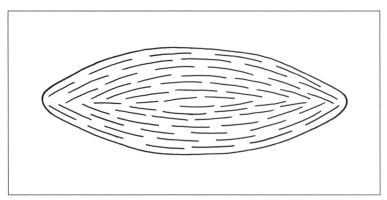

Figure 8-1   Fusulinids are important guide fossils for the Carboniferous.

lobe-finned, air-breathing fish that grew upwards of 10 feet long with large teeth and the predecessors of modern lungfish. An abundance of food swept up on the beaches during high tide might have enticed these fish to come ashore. Fierce competition in the ocean for a scarce food supply provided extraordinary evolutionary incentive for any animal that could find food on land.

Amphibious fish probably spent little time on shore because their primitive legs could not support their body weight for long periods, requiring them to return to the sea. Eventually, as their limbs strengthened, the amphibious fish wandered farther inland, where crustaceans and insects

Figure 8-2   Evolutionary progress from crossopterygian (top) to amphibious fish (middle) to amphibians (bottom).

were abundant. By the middle Devonian, they began to dominate the land and were especially attracted to the great Carboniferous swamps.

By about 335 million years ago, the amphibious fish had evolved into the earliest amphibians (Fig. 8-2). Some species had strong, toothy jaws and resembled giant salamanders, reaching 3 to 5 feet in length. Although the amphibians had well-developed legs for walking on dry land, they apparently spent most of their time in water and depended on nearby water sources to moisten their skins for respiration. They reproduced like fish, laying small eggs without protective membranes in water or moist places. After hatching, the juveniles lived an aquatic, fishlike life-style and breathed with gills. As they matured, the young amphibians metamorphosed into air breathing, four-limbed adults.

The early amphibians were slow and ungainly creatures, with weak legs that could hardly keep their squat bodies off the ground. Therefore, to succeed as hunters without the benefit of speed or agility, the amphibians developed a remarkable tongue that lashed out at insects and flicked them into their mouths. The adaptation was so successful, the amphibians rapidly populated the land.

The necessity of having to live a semiaquatic life-style led to the eventual downfall of the amphibians when the great swamps began to dry out toward the end of the Paleozoic. The void left by the amphibians was quickly filled by their cousins the reptiles, which were better suited for a life totally out of water and destined to become the greatest evolutionary success story the world has ever known.

THE GREAT COAL FORESTS

During the second half of the Paleozoic, the continents rose and sea levels dropped, causing the departure of the inland seas, which were replaced with immense swamps. About 315 million years ago, extensive forests grew in the great swamps. These regions formed a vast tropical belt that ran through the supercontinent Pangaea, which straddled the equator.

The lycopods, which ruled the ancient swamps, towering as high as 130 feet, were the first trees to develop true roots and leaves that were generally small. Branches were arranged in a spiral and spores were attached to modified leaves that became primitive cones. Their trunks were composed largely of bark, and for the most part lacked branches along the length of the tree, so the trees looked much like a forest of telephone poles. Only near the end of their lives did the lycopods sprout a small crown of limbs as they prepared for reproduction. Making their living among these trees were giant insects, huge millipedes, walking fish, and primitive amphibians.

For millions of years, the lycopods endured changes in sea level and climate that alternately drained and flooded the swamps. Then about 310

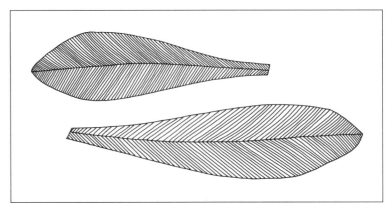

Figure 8-3 Fossil glossopteris leaves from the middle to late Paleozoic helped prove the theory of continental drift.

Figure 8-4 A fossil leaf from the Raton Formation near Trinidad, Colorado. Photo by W. T. Lee, courtesy of USGS

million years ago, the climate of the tropics became drier and most swamp-lands disappeared entirely. The climate change set off a wave of extinctions that wiped out virtually all lycopods. Today, they exist only as small grass-like plants in the tropics. Late in the Carboniferous, as the climate grew wetter and the swamps reemerged, weedy plants called tree ferns dominated the Paleozoic wetlands.

The second most diverse group of living plants were the true ferns. They range from the Devonian to the present but were particularly widespread in the Mesozoic and prospered well in the mild climates, whereas today they are restricted to the tropics. Some ancient ferns attained heights of present-day trees. The Permian seed fern glossopteris was especially significant. Its fossil leaves (Fig. 8-3) are prevalent on the continents that formed Gondwana but are lacking on the continents that comprised Laurasia, indicating that these two landmasses were in separate parts of the world divided by the Tethys Sea. This sea was wide in the east and narrow in the west, where land bridges aided the migration of plants and animals from one continent to the other.

Although terrestrial fossils are not nearly as abundant as marine fossils, primarily because land species do not fossilize well and fossil-bearing sediments are subjected to erosion, some environments like ancient swamps and marshes provide an abundance of plant and animal fossils. Well preserved, carbonized plant material is commonly found between easily-separated sediment layers (Fig. 8-4). Animals were also buried in the great coal swamps, where their bones were preserved and fossilized.

FOSSIL FUELS

The Carboniferous and Permian had the highest organic burial rates of any period in Earth history. Extensive forests and swamps grew successively on top of each other and continued to add to thick deposits of peat, which were buried under layers of sediment. The weight of the overlying strata and heat from the Earth's interior reduced the peat to about 5 percent of its original volume and metamorphosed it into lignite, as well as bituminous and anthracite coal.

The world's coal reserves far exceed all other fossil fuels combined and are sufficient to support large increases in consumption well into the next century. The amount of economically recoverable coal reserves is upwards of one trillion tons. The United States holds substantial reserves of coal (Fig. 8-5), which remain practically untouched. Since coal is the cheapest and most abundant fossil fuel, it will be a favorable alternate source of energy to replace petroleum when reserves run low.

Paleozoic sediments hold a large portion of the world's oil reserves, indicating a high degree of marine organic productivity during this time. The formation of oil and gas requires special geologic conditions, including

Figure 8-5 Open pit coal mining at the West Decker mine, Montana. Photo by P. F. Narten, courtesy of USGS

a sedimentary source of organic material, a porous rock to serve as a reservoir, and a confining structure to act as a trap. The source material is organic carbon in fine-grained, carbon-rich sediments. Porous and permeable sedimentary rocks such as sandstones and limestones serve as reservoir. Geologic structures created by folding or faulting of sedimentary layers trap or pool the oil and gas.

Most of the organic material that produces petroleum derives from microscopic organisms that originated primarily in the surface waters of the ocean and were concentrated in fine particulate matter on the seafloor. For organic material to become petroleum, either the rate of accumulation must be high or the oxygen level in the bottom water must be low so the material does not oxidize before burial under thick sedimentary layers. This is because oxidation causes decay, which destroys organic material. Therefore, areas with high rates of accumulation of sediments rich in organic material are the most favorable sites for the formation of oil-bearing rock.

After deep burial in a sedimentary basin, high temperatures and pressures generated in the Earth's interior chemically alter the organic material into hydrocarbons. If the hydrocarbons are overcooked, natural gas results. Oil is often associated with thick beds of salt. Because salt is lighter than the overlying sediments, it rises toward the surface, creating salt domes that help trap the oil.

Hydrocarbon volatiles (fluids and gases) along with seawater locked up in the sediments migrated upward through permeable rock layers and accumulated in traps formed by sedimentary structures that provide a barrier to further migration. In the absence of a cap rock, the volatiles continue rising to the surface and escape. Much petroleum has been lost in this manner, as well as by the destruction of the reservoirs by uplift and erosion of the confining structure. Several tens of millions to a few hundred million years are required to process organic material into oil, depending mainly on the temperature and pressure conditions within the sedimentary basin.

CARBONIFEROUS GLACIATION

During the latter part of the Carboniferous around 290 million years ago, Gondwana was in the south polar regions, where glacial centers expanded across the continents (Fig. 8-6). Rocks heavily grooved by the advancing glaciers show lines of ice flow away from the equator and toward the poles, which would not be possible if the continents were situated where they are today. Furthermore, the ice would have had to flow from the sea onto the land in many areas, which is highly unlikely. Instead, the southern continents drifted en masse over the South Pole, and massive ice sheets crossed the present continental boundaries.

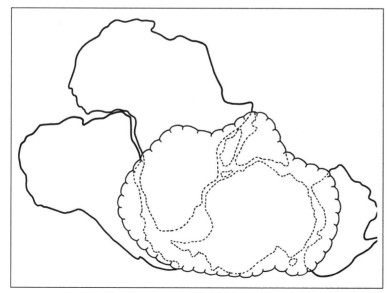

Figure 8-6   The extent of late Paleozoic glaciation in Gondwana.

Figure 8-7   A thick coal bed at Little Powder River coal field, Montana. Photo by Dobbin, courtesy of USGS

During the early part of the glacial epoch, the maximum glacial effects were in South America and South Africa. Later, the chief glacial centers moved to Australia and Antarctica, clearly showing that the southern continents that comprised Gondwana hovered over the South Pole. Continents residing near the poles are often the cause of extended periods of glaciation because land at higher latitudes has a high albedo (reflective quality) and a low heat capacity, encouraging the accumulation of ice.

Ice sheets covered large portions of east-central South America, South Africa, India, Australia, and Antarctica. In Australia, marine sediments were interbedded with glacial deposits and tillites were separated by seams of coal, indicating that periods of glaciation were interspersed with warm interglacial spells, when extensive forests grew. In South Africa, the Karroo Series, comprising a sequence of late Paleozoic tillites and coal beds, extended over an area of several thousand square miles and reached a total thickness of 20,000 feet. Among the coal beds, the best in Africa, are fossil glossopteris leaves, whose existence on the southern continents is among the best evidence for the theory of continental drift.

One cause of the ice age was the loss of atmospheric carbon dioxide. The burial of carbon dioxide in the crust might have been the key to the onset of all major ice ages since life evolved on Earth. A substantial carbon dioxide repository during the latter part of the Paleozoic were the great forests that spread over the land. Plants began to invade the land and extended to all parts of the world beginning about 450 million years ago. Lush forests that grew during the Carboniferous stored large quantities of carbon in their woody tissues. Burial under layers of sediment compacted the vegetative matter and converted it into thick seams of coal (Fig. 8-7). The reduction of the carbon dioxide content in the atmosphere severely weakened the greenhouse effect, causing the climate to cool.

The continental margins became less extensive and narrower, confining marine habitats to near-shore regions, which might have influenced the great extinction at the end of the Paleozoic. Land once covered with great coal swamps dried out as the climate grew colder. No major extinction event occurred during the widespread Carboniferous glaciation around 330 million years ago, however. The relatively low extinction rates were

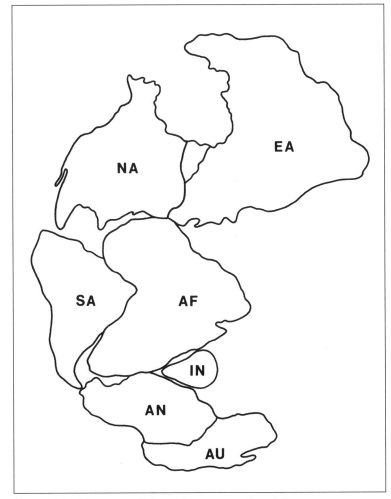

Figure 8-8   The upper Paleozoic supercontinent Pangaea.

probably due to a limited number of extinction-prone species following the late Devonian extinction.

PANGAEA

Between 360 and 270 million years ago, Gondwana and Laurasia converged into the supercontinent Pangaea (Fig. 8-8), meaning all lands. The massive continent had a total area of about 80 million square miles or 40 percent of the Earth's total surface area. It straddled the equator, extending practically from pole to pole, with an almost equal amount of land in both hemispheres, whereas today two-thirds of the continents lie north of the equator. A single great ocean called Panthalassa stretched uninterrupted across the planet, with the continents huddling to one side. Over the ensuing periods, smaller parcels of land continued to collide with the supercontinent until it reached its peak size about 210 million years ago.

The continental collisions crumpled the crust and pushed up huge masses of rock into several mountain belts throughout many parts of the world (Fig. 8-9). Volcanic eruptions were extensive due to frequent continental collisions. During times of highly active continental movements, volcanic activity increases, especially at midocean spreading ridges where new oceanic crust is created and at subduction zones where old oceanic

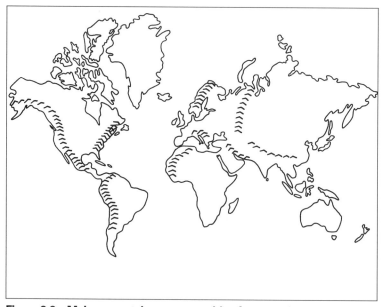

Figure 8-9   Major mountain ranges resulting from continental collisions.

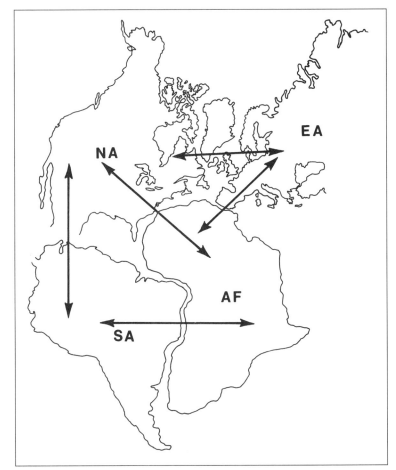

Figure 8-10   The effect of geography on the migration of species.

crust is destroyed. The amount of volcanism affects the rate of mountain building and the composition of the atmosphere, which affects the climate.

When Gondwana linked with Laurasia to form Pangaea, the collision raised the Appalachian and Ouachita mountains. Simultaneously, Laurasia connected with Siberia, thrusting up the Ural Mountains. The continued clashing of island arcs with North America resulted in an episode of mountain building in Nevada called the Sonoma orogeny, which coincided with the complete assembly of Pangaea 250 million years ago.

The sediments in the Tethys Sea separating Gondwana and Laurasia buckled and uplifted into various mountain belts, including the ancestral Hercynian Mountains of southern Europe. As the continents rose higher and the ocean basins dropped lower, the land became dryer and the climate grew colder, especially in the southernmost lands, which were covered with glacial ice. All known episodes of glaciation occurred during times of lowered sea levels. The changes in the shapes of the ocean basins greatly influenced the course of ocean currents, which in turn had a pronounced affect on the climate.

The closing of the Tethys Sea eliminated a major barrier to the migration of species from one continent to another, and they dispersed to all parts of the world (Fig. 8-10). When all continents combined into Pangaea, plant and animal life experienced a proliferation of species in the ocean as well as on land. The formation of Pangaea marked a major turning point in the evolution of life, during which the reptiles emerged as the dominant species, conquering land, sea, and sky.

A continuous shallow-water margin ran around the entire perimeter of Pangaea, and no major physical barriers hampered the dispersal of marine life. Furthermore, the seas were largely confined to the ocean basins, leaving the continental shelves mostly exposed. The continental margins were less extensive and narrower, confining marine habitats to the near-shore regions. Consequently, habitat area for shallow-water marine organisms was very limited, which accounts for the low species diversity. As a result, marine biotas were more widespread but contained comparatively fewer species.

In the northern latitudes, thick forests of primitive conifers, horsetails, and club mosses that grew as tall as 30 feet dominated the mountainous landscape. Much of the interior probably resembled a grassless version of the steppes of central Asia, where temperatures varied from very hot in

Figure 8-11   Moschops were large herbivorous reptiles that traveled in vast herds.

summer to very cold in winter. Since grasses would not appear for well over 100 million years, the scrubby landscape was dotted with bamboolike horsetails and bushy clumps of now-extinct seed ferns that resembled present-day tree ferns.

Browsing on the seed ferns were herds of moschops (Fig. 8-11), 16-foot reptiles with thick skulls adapted for butting during mating season, a tactic similar to the behavior of modern herd animals. They were probably preyed upon by packs of lycaenops, which were reptiles with doglike bodies and long canine teeth projecting from their mouths. Mammal-like reptiles called dicynodonts also had two caninelike tusks and fed on small animals along riverbanks. A 2-foot-long amphibian with armadillolike plates rooted in the soil for worms and snails. Small reptiles probably ate insects like modern lizards do.

The Pangaean climate was one of extremes, with the northern and southern regions colder than Siberia and the interior deserts hotter than the Sahara, and almost devoid of life. The massing of continents together created an overall climate that was hotter, drier, and more seasonal than at any other time in geologic history. These conditions might have led to the mass extinction of primarily terrestrial animals some 30 million years after most of Pangaea was assembled. Pangaea remained intact for another 40 million years, after which it rifted apart into the present continents.

9

PERMIAN REPTILES

The Permian period from 280 to 250 million years ago was named for a well-exposed sequence of marine rocks and terrestrial redbeds on the western side of the Ural Mountains in the Russian district of Perm. Rocks of Permian age are distinct in western North America, particularly in Nevada, Utah, and Texas. Important reserves of oil and natural gas reside in the Permian Basin of Texas and Oklahoma. Extensive coal deposits of Permian age exist in Asia, Africa, and Australia.

During the Permian, all major continents combined into the supercontinent Pangaea, where widespread mountain building and extensive volcanism were prevalent. The interior of Pangaea was largely desert, causing the decline of the amphibians and the rise of the reptiles. At the end of the Permian, perhaps the greatest extinction the Earth has ever known eliminated over 95 percent of all species, paving the way for the ascension of the dinosaurs.

THE AGE OF REPTILES

The age of the reptiles, which began in the Permian and lasted 200 million years, witnessed the evolution of some 20 orders of reptilian families.

Figure 9-1 Evolutionary stages in the development of the reptile, from the amphibious fish (top), to the amphibians (middle), to the reptiles (bottom).

Amphibians, which were prominent in the Carboniferous, declined considerably in the Permian because of a preference for life in water. When the Carboniferous swamps dried out and were largely replaced with deserts, the amphibians gave way to their cousins the reptiles (Fig. 9-1), which were well adapted to drier climates. In the latter part of the Permian, the reptiles succeeded the amphibians and became the dominant land-dwelling animals of the Mesozoic era. The generally warm climate of the era was also advantageous to the reptiles and aided them in colonizing the land.

The increase in the number of reptilian fossil footprints in Carboniferous and Permian sediments shows the rise of the reptiles, at the expense of the amphibians. The success of the reptiles was largely due to their more efficient mode of locomotion. The reptiles were also better suited to a full-time life on dry land, whereas the amphibians were dependent on a local source of water for moistening their skins and reproduction.

The reptilian foot was a major improvement over that of the amphibian, with changes in the form of the digits, the addition of a thumblike fifth digit, and the appearance of claws. In some reptiles, the tracks narrowed and the stride lengthened. Others maintained a four-footed walking gait and ran reared up on their hind legs. Although most reptiles walked or ran on all fours, by the late Permian some smaller reptiles often stood on their hind legs when they required speed and agility. The body pivoted at the hips and a long tail counterbalanced the nearly erect trunk. This stance freed the forelimbs for attacking small prey and completing other useful tasks.

Reptiles have scales that retain the animal's bodily fluids, whereas amphibians have a permeable skin that must be moistened frequently. Another major advancement over the amphibians was the reptile's mode of reproduction. Like fish, amphibians laid their eggs in water, and after hatching, the young fended for themselves, often becoming prey for predators. The reptile's eggs have hard, watertight shells so they can be laid on dry land. Parents protected their young, which gave them a better chance of survival, contributing to the reptile's great success in populating the land.

Like fish and amphibians, reptiles are cold-blooded, a term that is misleading since they draw heat from the environment. Therefore, the blood of a reptile sunning on a rock can actually be warmer than that of a warm-blooded mammal. An ancient reptile called dimetrodon (Fig. 9-2) had a large sail along its back, apparently to regulate its body temperature by absorbing sunlight during cold weather and radiating excess body heat

Figure 9-2 Pelycosaurs, like dimetrodon, had large dorsal sails to regulate their body temperature.

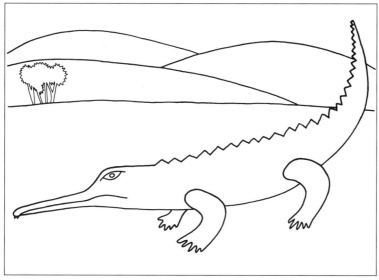

Figure 9-3 Phytosaurs were predatory reptiles that resembled crocodiles but were not closely related to them.

when the weather was hot. A high body temperature is as important to a reptile as it is to a mammal to achieve maximum metabolic efficiency. In cold mornings, reptiles are sluggish and vulnerable to predators. They bask in the sun until their bodies warm and their metabolism can operate at peak performance.

Reptiles require only about one-tenth the amount of food mammals need to survive because mammals use most of their calories to maintain a high body temperature. Consequently, reptiles can live in deserts and other desolate places and flourish on small quantities of vegetation that would quickly starve a mammal of the same size. The generally warm climate of the Mesozoic was very advantageous to the reptiles and aided them in colonizing the land, whereas the amphibians, which avoided direct sunlight and were relatively cold and slow-moving, were at a disadvantage.

Perhaps the strangest reptile that ever lived was tanystropheus, dubbed the "giraffe-neck saurian." The animal measured as much as 15 feet from head to tail and is famous for its absurdly long neck, which was more than twice the length of the trunk. As it matured, its neck grew at a much faster rate than the rest of its body. The reptile was most likely aquatic because it could not possibly have supported the weight of its neck while on land. Tanystropheus probably used its grossly elongated neck for scavenging bottom sediments for food.

The phytosaurs (Fig. 9-3) were large, heavily armored, predatory reptiles with sharp teeth. They resembled crocodiles with their elongated snouts, short legs, and long tails, but were not closely related to them. They evolved from the thecodonts, which also gave rise to the crocodiles and dinosaurs. They thrived in the late Triassic, evolving quite rapidly, but apparently did not survive beyond the end of the period.

Near the close of the Triassic, when reptiles were the leading form of animal life, occupying land, sea, and air, a remarkable reptilian group called the crocodilians appeared in the fossil record. This group had originated on Gondwana, and was composed of the alligators with a blunt head, the crocodiles with an elongated head, and the gavials with an extremely narrow head. Members of this group adapted to life on dry land, a semiaquatic life, or an entirely aquatic life with a sharklike tail, a streamlined head, and legs modified into swimming paddles. A fossil of a gavial-like monster from the lower Cretaceous in Niger, West Africa, measured about 35 feet long.

The crocodilians diversified considerably over the past 200 million years, spreading to all parts of the world and adapting to a wide variety of habitats. Crocodile fossils found in the high latitudes of North America (Fig. 9-4) indicate a warm climate during the Mesozoic. They belong to the subclass Archosauria, which literally means ruling reptiles, along with dinosaurs and pterosaurs and are the only surviving members to escape the "great dying" at the end of the Cretaceous.

Figure 9-4 The Chinle Formation, Uinta Range, Utah, where crocodile fossils are found.

Figure 9-5 Many therapsids were large predatory mammal-like reptiles.

MAMMAL-LIKE REPTILES

The pelycosaurs were the first animals to depart from the basic reptilian stock some 300 million years ago. They were distinguished from other reptiles by their large size and varied diet, and the earliest predators were capable of killing relatively large prey, including other reptiles. A pelycosaur called dimetrodon grew to a length of about 11 feet. It had a large dorsal sail composed of webs of membrane well supplied with blood, stretching across bony protruding spines and probably used for temperature control. When the animal was cold, it turned its body broadside to the sun to absorb more sunlight. When the animal was hot, it sought out a shaded area or exposed itself to the wind. This appendage might have been a crude forerunner of the temperature control system in mammals.

As the climate warmed, the pelycosaurs lost their sails and perhaps gained some degree of internal thermal control. They thrived for about 50 million years and then gave way to their descendants, the mammal-like reptiles called the therapsids (Fig. 9-5). The first therapsids retained many characteristics of the pelycosaurs, but their legs were well adapted for much higher running speeds. They ranged in size from as small as a mouse to as large as a hippopotamus.

The early members invaded the southern continents at the beginning of the Permian when those lands were recovering from the Carboniferous glaciation, suggesting the animals were warm-blooded enough to withstand the cold. They probably had undergone some physiological adaptations to enable them to feed and travel through the snows of the cold winters. They were apparently too large to hibernate, as shown by the lack of growth rings in their bones, an indicator similar to the tree rings that mark alternating seasons of growth. The development of fur appeared in the more advanced therapsids, as they migrated into colder climates. Therapsids might also have operated at lower body temperatures than most living mammals to conserve energy. The family of mammal-like reptiles clearly shows a transition from reptile to mammal. Mammals evolved from the mammal-like reptiles over a period of more than 100 million years, during which time the animals adapted so as to function better in a terrestrial environment. Teeth evolved from simple cones that were replaced repeatedly during the animal's lifetime to more complex shapes that were replaced only once. However, the mammalian jaw and other parts of the skull still shared many similarities with reptiles.

The advantages of being warm-blooded are tremendous, and a stable body temperature finely tuned to operate within a narrow thermal range provides a high rate of metabolism independent of the outside temperature. Therefore, the work output of leg muscles, heart, and lungs increases enormously, giving mammals the ability to outperform and outendure reptiles. The principle of heat loss, by which a large body radiates more thermal energy than a small one, applies to large reptiles as well as mammals. In addition, mammals have a coat of insulation, including an

Figure 9-6   The Blue Ridge Mountains in the Appalachians, Avery County, North Carolina. Photo by D. A. Brobst, courtesy of USGS

outer layer of fat and fur, to prevent the escape of body heat during cold weather.

The therapsids appear to have reproduced like reptiles, by laying eggs. They might have protected and incubated the eggs and fed their young. This in turn might have resulted in longer egg retention in the female and given rise to live births. The therapsids dominated animal life for more than 40 million years until the middle Triassic, and then for unknown reasons they lost out to the dinosaurs. From then on, primitive mammals were relegated to the role of a shrewlike nocturnal hunter of insects until the dinosaurs finally became extinct.

THE APPALACHIAN OROGENY

Perhaps the most impressive landforms on Earth are the mountain ranges, created by the forces of uplift and erosion. Paleozoic continental collisions crumpled the crust, pushing up huge masses of rock into several mountain chains throughout many parts of the world.

The Appalachian Mountains (Fig. 9-6), extending some 2,000 miles from central Alabama to Newfoundland, were upraised during continental collisions between North America, Eurasia, and Africa in the late Paleozoic, from about 350 million to 250 million years ago during the construction of Pangaea. The southern Appalachians are underlain with more than 10

Figure 9-7 A wax model illustrating mountain uplift. Photo by J. K. Hillers, courtesy of USGS

miles of flat-lying sedimentary and metamorphic rocks, whereas the surface rocks were highly deformed by the collision (Fig. 9-7).

This type of formation suggests that these mountains were the product of thrust faulting, involving crustal material carried horizontally for great distances. The sedimentary strata rode westward on top of Precambrian metamorphic rocks and folded over, buckling the crust into a series of ridges and valleys. The existence of sedimentary layers beneath the core of the Appalachians suggests that thrusting involving basement rocks is responsible for the formation of all mountain belts, possibly since the process of plate tectonics began. The shoving and stacking of thrust sheets during continental collision also might have been a major mechanism in the continued growth of the continents.

The Mauritanide mountain range in Western Africa is the counterpart, the "other side," of the Appalachians. It is characterized by a series of belts running east to west that are similar in many respects to the Appalachian belts. The eastern parts of the range comprise sedimentary strata partially covered with metamorphic rocks that have overridden the sediments from the west along thrust faults. Older metamorphic rocks resembling those of the southern Appalachians lie westward of this region, while a coastal plain of younger horizontal rocks covers the rest. Furthermore, a period of metamorphism and thrusting similar to the formation of the Appalachians occurred prior to the opening of the Atlantic. In this respect, the two mountain ranges are practically mirror images of each other.

This episode of mountain building also uplifted the Hercynian Mountains in Europe, which extended from England to Ireland and continued through France and Germany. The folding and faulting was accompanied by large-scale igneous activity in England and on the European continent. The Ural Mountains were similarly formed by a collision between the Siberian and Russian continental shields. The Transantarctic Range, comprising great belts of folded rocks, formed when two plates came together to create the continent of Antarctica. Prior to the end of the Permian, the younger parts of West Antarctica had not yet formed, and only East Antarctica was present.

LATE PALEOZOIC GLACIATION

The continents of Africa, South America, Australia, India, and Antarctica were glaciated in the late Paleozoic, around 270 million years ago, as evidenced by glacial deposits and striations in ancient rocks. The lines of ice flow were away from the equator and toward the poles. Therefore, the continents must have joined in such a manner that ice sheets moved across a single landmass, radiating outward from a glacial center over the South Pole.

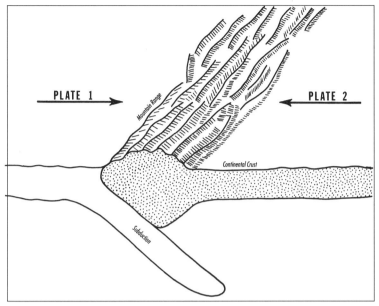

Figure 9-8 **The formation of mountains by the convergence of two lithospheric plates.**

The Late Paleozoic was a period of extensive mountain building, which raised massive chunks of crust to higher elevations, where glaciers are nurtured in the cold, thin air. Glaciers also might have formed at lower latitudes when lands were elevated during continental collisions (Fig. 9-8). When Gondwana and Laurasia combined into Pangaea, the continental collisions crumpled the crust and pushed up huge blocks into several mountain chains throughout many parts of the world. Generally speaking, with high mountains come low temperatures and increased precipitation, a situation that maintains glaciers in the high altitudes.

Besides folded mountain belts, volcanoes were prevalent, and unusually long periods of volcanic activity blocked out the sun with clouds of volcanic dust and gases, thereby significantly lowering surface temperatures. As the continents rose higher, the ocean basins dropped lower. The change in shape of the ocean basins greatly affected the course of ocean currents (Table 9-1), which in turn had a profound effect on the climate. All known episodes of glaciation occurred at times when sea levels should have been low, although not all mass extinctions were associated with lowered sea levels.

The continental margins became less extensive and narrower, confining marine habitats to near-shore areas. Such an occurrence might have had a major influence on the great extinction at the end of the Paleozoic era. During this time, land once covered with great coal swamps completely

TABLE 9–1   HISTORY OF THE DEEP CIRCULATION IN THE OCEAN

Age (millions of years ago)	Event
3	An ice age overwhelms the Northern Hemisphere.
3–5	Arctic glaciation begins.
15	The Drake Passage is open; the circum-Antarctic current is formed. Major sea ice forms around Antarctica, which is glaciated, making it the first major glaciation of the Modern Ice Age. The Antarctic bottom water forms. The snow limit rises.
25	The Drake Passage between South America and Antarctica begins to open.
25–35	A stable situation exists with possible partial circulation around Antarctica. The equatorial circulation is interrupted between the Mediterranean Sea and the Far East.
35–40	The equatorial seaway begins to close. There is a sharp cooling of the surface and of the deep water in the south. The Antarctic glaciers reach the sea with glacial debris.The seaway between Australia and Antarctica opens. Cooler bottom water flows north and flushes the ocean. The snow limit drops sharply.
>50	The ocean could flow freely around the world at the equator. Rather uniform climate and warm ocean even near the poles. Deep water in the ocean is much warmer than it is today. Only alpine glaciers on Antarctica.

dried out as the climate grew colder, culminating in the deaths of multitudes of species.

MASS EXTINCTION

Throughout the Earth's history, massive numbers of species have vanished in several short periods (Fig. 9-9). During geologically brief intervals of perhaps a few million years, mass extinctions in the ocean have eliminated

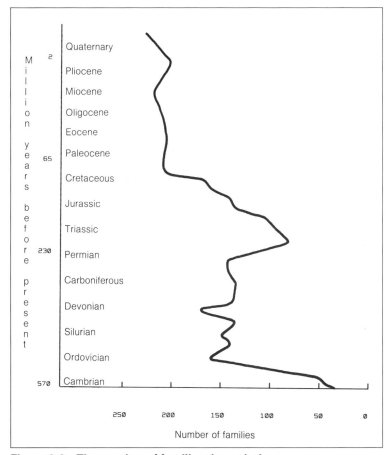

Figure 9-9   The number of families through time.

half or more of the existing families of plants and animals. Devastations of this magnitude are generally due to radical global changes in the environment. Drastic changes in environmental limiting factors, including temperature and living space on the ocean floor, determined the distribution and abundance of species in the sea.

Many episodes of extinction coincided with periods of glaciation, and global cooling had a major effect on life. The living space of warmth-loving species was restricted to the tropics. Species trapped in confined waterways, unable to move to warmer seas, were particularly hard hit. Furthermore, the accumulation of glacial ice in the polar regions lowered sea levels, thereby reducing shallow water shelf areas, which limited the amount of habitat and consequently the number of species.

Ocean temperature is by far the most important factor limiting the geographic distribution of marine species, and climatic cooling is the primary culprit behind most extinctions in the sea. Species unable to migrate to warmer regions or adapt to colder conditions are usually the most adversely affected. This is especially true for tropical faunas that can tolerate only a narrow range of temperatures and have nowhere to migrate. Since lowered temperatures also slow down the rate of chemical reactions, biological activity during a major glacial event should function at a lower energy state, which could affect the rate of evolution and species diversity.

The greatest extinction event took place when the Permian ended 250 million years ago. The extinction was particularly devastating to Permian marine fauna (Fig. 9-10). Half the families of marine organisms, including more than 95 percent of all known species, abruptly disappeared. In effect, the extinction left the world almost as devoid of life forms at the end of the Paleozoic as at the beginning. The extinction followed on the heels of a late Permian glaciation, when thick ice sheets blanketed much of the Earth, significantly lowering ocean temperatures.

Corals, which require warm, shallow water, were affected the worst, as evidenced by the lack of coral reefs in the early Triassic. Another group of animals that disappeared at this time were the fusulinids, which were formanifers shaped like a grain of wheat and populated the shallow seas of

Figure 9-10 Marine flora and fauna of the Permian. Courtesy of Field Museum of Natural History, Chicago

the world for about 80 million years, during which their shells accumulated into vast deposits of limestone.

The crinoids and brachiopods, which had their heyday in the Paleozoic, were relegated to minor roles during the succeeding Mesozoic. The spiny brachiopods that were plentiful in the late Paleozoic seas vanished without leaving any descendants. The trilobites, which were extremely successful during the Paleozoic, completely died out at the end of the era. A variety of other crustaceans, including shrimps, crabs, crayfish, and lobsters, occupied the habitats vacated by the trilobites. On land, 75 percent of the amphibian families and over 80 percent of the reptilian families disappeared.

Whatever the agents of biological stress were—climatic changes, shifts in ocean currents, shallowing seas, or disruptions in food chains—the ability of the biosphere to resist them varied in different parts of the world.

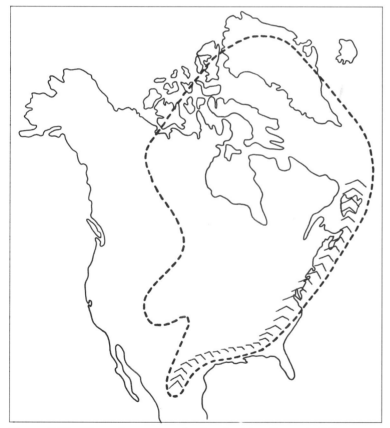

Figure 9-11 The paleogeography (ancient landforms) of the upper Paleozoic in North America.

But one very consistent pattern in mass extinctions was that, although each event typically affected different suites of organisms, tropical biotas, which contain the highest number of species, were almost always the hardest hit.

When all continents had converged into Pangaea by the end of the Permian around 250 million years ago, the change in geography spurred a great proliferation of plant and animal life on land and in the sea. The formation of Pangaea marked a major turning point in the evolution of life, during which the reptiles emerged as the dominant species.

The Pangaean climate appears to have been equable and warm throughout most of the year. However, much of the interior of Pangaea was desert, where temperatures fluctuated wildly from season to season, with scorching hot summers and freezing cold winters. These climate conditions might have contributed to the widespread extinction of land-based species during the late Paleozoic. It also explains why the reptiles, which adapt readily to hot, dry climates, replaced the amphibians as the dominant land animals.

The sea level lowered during the formation of Pangaea and drained the interiors of the continents (Fig. 9-11). The drop in sea level caused the inland seas to retreat, producing a continuous, narrow continental margin around the supercontinent. This in turn reduced the amount of shoreline, which radically limited the marine habitat area. Moreover, unstable near-shore conditions resulted in an unreliable food supply. Many species unable to cope with the limited living space and food supply died out in tragically large numbers, paving the way for the ascension of entirely new species.

10

TRIASSIC DINOSAURS

The Triassic period, which marks the start of the Mesozoic era, ran from 250 to 210 million years ago and was named for a sequence of redbed and limestone strata in central Germany. In North America, continental sediments and redbeds add to the rugged beauty of Utah, Wyoming, and Colorado. In Arizona's Petrified Forest lie the fossilized remains of primitive Triassic-age conifers (Fig. 10-1) that once flourished in the upland regions. At the end of the period, the supercontinent Pangaea started rifting apart into the present continents, and massive amounts of basalt spilled onto the landscape.

During the late Triassic, large families of terrestrial animals died off in record numbers. The mass extinction spanned a period of less than a million years but was responsible for killing nearly half the reptile families. The dying-out of species forever changed the character of life on Earth and initiated the rise of the dinosaurs, one of biology's greatest success stories.

THE AGE OF DINOSAURS

At the beginning of the Mesozoic era, the continents consolidated into a supercontinent, at the era's midpoint they began to rift apart, and at its end

Figure 10-1   Petrified trees at the Petrified Forest National Monument, Apache County, Arizona. Photo by Darton, courtesy of USGS

they were well on the way to their present locations. The breakup of Pangaea created three new bodies of water that included the Atlantic, Arctic, and Indian oceans. The climate was exceptionally mild for an unusually long time. The reptiles especially prospered during these extraordinary conditions. Besides conquering the land, some species went to sea and others took to the air, occupying nearly every corner of the globe.

Early in the Triassic, the Earth was recovering from a major ice age and an extinction event that took the lives of over 95 percent of all species. The early Mesozoic marked a rebirth of life, and 450 new families of plants and animals came into existence. But instead of inventing entirely new body plans as during the Cambrian explosion at the beginning of the Paleozoic, species developed new variations on already established themes. Therefore, fewer experimental organisms arose, and many lines of today's species evolved. Several major groups of terrestrial vertebrates made their debut, including the ancestors of dinosaurs, modern reptiles, mammals, and the predecessors of birds. The true birds did not appear in the fossil record for another 50 million years.

Dinosaurs arose to become the dominant terrestrial species for the next 150 million years. They suppressed the rise of other creatures, including the mammals, which were then tiny and inconsequential. The amphibians continued to decline during the Mesozoic, with all large, flat-headed species becoming extinct early in the Triassic. The group thereafter was represented by the more familiar toads, frogs, and salamanders. Although the amphibians did not achieve complete dominion over the land, their descendants the reptiles became the undisputed rulers of the world.

The oldest dinosaurs originated on the southern continent of Gondwana when the last glaciers of the great Permian ice age were departing and the region was still recovering from the cold conditions. About 230 million years ago, when mammal-like reptiles dominated the land, dinosaurs represented only a minor percentage of all animals. Several reptile species living at the time of the early dinosaurs still far outweighed them. However,

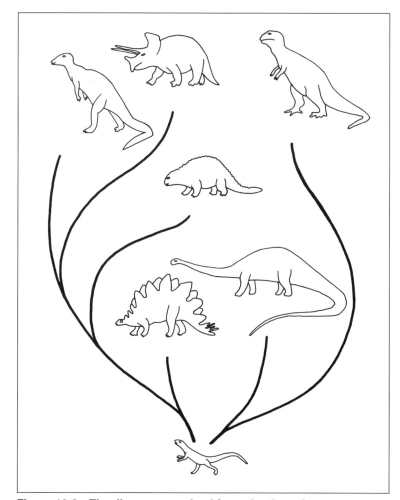

Figure 10-2   The dinosaurs evolved from the thecodonts.

Figure 10-3   The small plant eater camptosaur was an ancestor of many later dinosaurs. Courtesy of National Park Service

in just 10 million years, dinosaurs became the dominant species, evolving from moderate-sized animals less than 20 feet long to become the giants for which they are famous.

The dinosaurs descended from the thecodonts (Fig. 10-2), the apparent common ancestors of crocodiles and birds. The earliest thecodonts were small to medium-size predators that lived during the Permian-Triassic transition. One group of thecodonts took to the water and became large fish eaters. They included the phytosaurs, which died out in the Triassic, and the crocodilians, which remain successful today. Pterosaurs were also descendants of the Triassic thecodonts. The appearance of featherlike scales ostensively used for insulation suggests that thecodonts were also the ancestors of birds. By the end of the Triassic, the dinosaurs replaced the thecodonts as the dominant terrestrial vertebrates.

Dinosaurs are classified as sauropods, which were long-necked herbivores, or as carnosaurs, which were bipedal carnivores that possibly hunted sauropods in packs. The camptosaur (Fig. 10-3), ancestor of many later dinosaur species, was a herbivore up to 25 feet long. Not all dinosaurs were giants, however, and many were no larger than most modern mammals. Protoceratops and ankylosaurs (Fig. 10-4) were very common and ranged over wide areas like modern-day sheep. The smallest known dinosaur footprints are about the size of a penny. The smaller dinosaurs had hollow

Figure 10-4 **Ankylosaurs were common herbivorous dinosaurs.**

bones similar to those of birds. Some had long, slender hind legs, long delicate forelimbs, and a long neck, and if not for a lengthy tail their skeletons would closely resemble those of modern ostriches.

Many early small dinosaurs reared up on their hind legs and were the first animals to establish a successful permanent two-legged stance.

Figure 10-5 **Apathosaurs were among the largest dinosaurs.**

Bipedalism increased speed and agility and freed the forelimbs for foraging and other tasks. The back legs and hip supported the entire weight of the animal, while a large tail counterbalanced the upper portions of the body, and the dinosaur walked much like birds. Therefore, dinosaurs are classified as ornithischians with a birdlike pelvis or as saurischians with a lizard-like pelvis. The ornithischians probably arose from the same group of thecodont reptiles that gave rise to crocodiles and birds. Indeed, birds are the only living relatives of dinosaurs, and the skeletons of many small dinosaurs resemble those of birds.

Some bipedal dinosaurs later reverted to a four-footed stance probably because of their increased weight. They eventually evolved into gigantic long-tailed, long-necked sauropods such as the apathosaurus (Fig. 10-5), formerly called brontosaurus. Others, like tyrannosaurus rex, possibly the greatest land carnivore that ever lived, maintained a two-legged stance with powerful hind legs, a muscular tail for counterbalance, and arms shortened to almost useless appendages.

Dinosaur tracks are the most impressive of all fossil footprints because the great weight of many species caused deep indentations to be left in the ground (Fig. 10-6). Their footprints exist in relative abundance in terrestrial sediments of Mesozoic age throughout the world. The study of dinosaur tracks suggests that some species were highly gregarious, gathering in groups. Large carnivores like tyrannosaurus rex were swift, agile predators that could sprint up to 45 miles per hour.

Figure 10-6 Dinosaur trackways in the Chacarilla Formation, Tarapaca Province, Chile. Photo by R. J. Dingman, courtesy of USGS

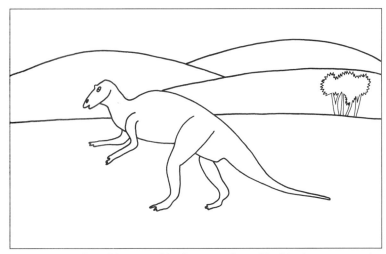

Figure 10-7   Fossil bones of hadrasaurs found in Alaska suggest the dinosaur might have migrated in immense herds when the climate grew cold.

The giant herbivores might have traveled in great herds, with the largest adults in the lead and the juveniles placed in the center for protection. Duck-billed hadrasaurs (Fig. 10-7), among the most successful of all dinosaur groups, were up to 15 feet tall and lived in the Arctic regions, where they either adapted to the cold and dark or migrated in large herds over long distances to warmer climates. Triceratops (Fig. 10-8), whose vast herds

Figure 10-8   Vast herds of triceratops roamed all parts of the world toward the end of the Cretaceous period.

roamed the entire globe toward the end of the Cretaceous and were among the last to go during the dinosaur extinction, might have contributed to the decline of other species of dinosaurs possibly due to extensive habitat destruction or the spread of diseases.

Females of some dinosaur species might have given live births. Many nurtured and fiercely protected their offspring until they could fend for themselves, allowing larger numbers to mature into adulthood, thus ensuring the continuation of the species. The parents might have brought food to their young and regurgitated seeds and berries as do modern birds. This parental care for infants suggests strong social bonds and might explain why dinosaurs were so successful for so long.

Some dinosaurs might have developed complex mating rituals. Besides regulating its body temperature, the large sail on dimetrodon's back might have been used to attract females. Other dinosaurs might have sported elaborate head gear for much the same purpose. The carnivores were cunning and aggressive creatures that charged at their prey with speed and agility. The cranial capacity of some carnivores indicates they possessed relatively large brains and were fairly intelligent, able to react to a variety of environmental pressures. The velociraptors with their sharp claws and powerful jaws were among the most vicious hunters, whose voracious appetites suggest they were warm-blooded.

Other dinosaur species might have acquired a certain degree of temperature control like mammals and birds. Dinosaurs descended from the thecodonts, the same common ancestors of warm-blooded birds, the distant living relatives of the dinosaurs. An argument in favor of warm-blooded dinosaurs contends that the skeletons of smaller, lighter species bear many resemblances to those of birds. Evidence for rapid juvenile growth, which is common among mammals, also exists in the bones of some dinosaur species, possibly providing another sign of warm-bloodedness.

When the dinosaur age began, the climates of southern Africa and the tip of South America where the early dinosaurs roamed experienced cold winters, during which large cold-blooded animals could not have survived without migrating to warmer regions. The stamina needed for such long-distance migration would have required sustained energy levels that only warm-blooded bodies could provide. Warm-blooded animals mature more rapidly than cold-blooded animals, which continue growing steadily until death. A comparison among the bones of dinosaurs, crocodiles, and birds, all of which had a common ancestor, shows a similarity between bird and dinosaur bones—another sign of possible warm-bloodedness. Furthermore, dinosaur bones had a higher blood vessel density than those of living mammals. Some dinosaur skulls show signs of sinus membranes, which exist only in warm-blooded animals.

Several dinosaur species appear to have been swift and agile, requiring a high rate of metabolism that only a warm-blooded body can provide. The complex social behavior of dinosaurs appears to be an evolutionary advancement that results from being warm-blooded. Even the females of some species might have produced live births like mammals. Yet at the end of the Cretaceous, when the climate supposedly grew colder, the warm-blooded mammals survived while the dinosaurs did not.

The period between the end of the Triassic and the beginning of the Jurassic was one of the most exciting times in the history of land vertebrates. In the late Triassic, all the continents were joined only at their western ends with Laurasia in the north and Gondwana in the south. Nonetheless, animal life on Laurasia was becoming distinct from that on Gondwana. A land bridge connecting Laurasia to the Indochina microcontinent might have been the last link, enabling the migration of animals when the two blocks collided at the close of the Triassic.

At the end of the Triassic, about 210 million years ago, a huge meteorite slammed into the Earth, creating the 60-mile-wide Manicouagan impact structure in Quebec, Canada. The gigantic explosion appears to have coincided with a mass extinction over a period of less than a million years that killed off 20 percent or more of all families of animals, including nearly half the reptile families. In the ocean, ammonoids and bivalves were decimated and the conodonts, fossil skeletons of a leechlike animal, completely disappeared. The extinction forever changed the character of life on Earth and paved the way for the rise of the dinosaurs.

Almost all modern animal groups, including amphibians, reptiles, and mammals, made their debut on the evolutionary stage at this time. This was also the time when the dinosaurs achieved dominance over the Earth—and held their ground for the next 150 million years. After this, theoretically, another large meteorite struck the Earth with the explosive force of 100 trillion tons of dynamite, turning the planet into an inhospitable world. In this manner, the dinosaurs might have been both created and destroyed by meteorites.

THE TETHYAN FAUNA

At the beginning of the Triassic, the Tethys Sea was a huge bay separating the northern and southern arms of Pangaea, which took the shape of a gigantic letter C that straddled the equator (Fig. 10-9). Between the late Paleozoic and middle Cenozoic, the Tethys was a broad tropical seaway that extended from western Europe to southeast Asia and harbored diverse and abundant shallow-water marine life. A great circumglobal ocean current that distributed heat to all parts of the world maintained warm climatic conditions. The energetic climate eroded the high mountain ranges of North America and Europe down to the level of the prevailing plain.

Early in the Triassic, ocean temperatures probably remained cool after the late Permian ice age. Marine invertebrates that managed to escape

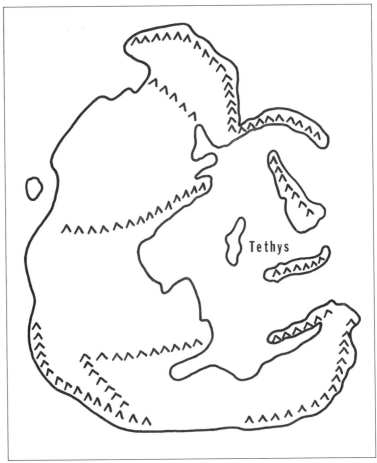

Figure 10-9   The Tethys Sea nestled between north and south Pangaea.

extinction lived in a narrow margin near the equator. Corals, which require warm, shallow water for survival, were particularly hard hit, as evidenced by the lack of coral reefs at the beginning of the Triassic. When the great glaciers melted and the seas began to warm, reef-building became intense in the Tethys Sea, with thick deposits of limestone and dolomite laid down by prolific lime-secreting organisms.

The mollusks appear to have weathered the hard times of the late Permian extinction quite well and continued on to become the most important shelled invertebrates of the Mesozoic seas, with some 60,000 distinct species living today. The warm climate of the Mesozoic influenced the growth of giant animals in the ocean as well as on land. Giant clams grew to 3 feet wide, giant squids were upwards of 65 feet long and weighed over a ton, and crinoids reached 60 feet in length.

The cephalopods were extremely spectacular and diversified to become the most successful marine invertebrates of the Mesozoic seas. The coiled-shell ammonoids, which evolved in the early Devonian period 395 million years ago, grew as much as 7 feet across. They traveled by jet propulsion, using neutral buoyancy to maintain depth, which contributed to their great success. However, of the 25 families of widely ranging ammonoids in the late Triassic, all but one or two became extinct at the end of the period. Those species that escaped extinction eventually evolved into scores of ammonite families in the Jurassic and Cretaceous.

Among the marine vertebrates, fish progressed into more modern forms. The sharks regained ground lost from the great Permian extinction and continued to become the successful predators they are today. The sea serpentlike plesiosaurs, the seacowlike placodonts, and the dolphinlike ichthyosaurs (Fig. 10-10) were reptiles that returned to the sea, where they achieved great evolutionary success. The placodonts were a group of short, stout marine reptiles with large, flattened teeth, which they probably used to feed primarily on bivalves and other mollusks. Many other reptilian species, including lizards and turtles that were quite primitive in the Triassic, also went to sea. However, only the smallest turtles made it past the extinction at the end of the Cretaceous.

During the final stages of the Cretaceous, when the seas drained from the land as the level of the ocean dropped, the temperatures in the Tethys Sea began to fall. As the continents drifted poleward during the last 100 million

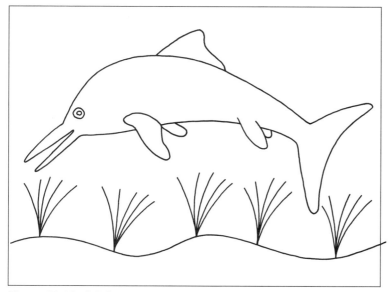

Figure 10-10   Ichthyosaurs were large marine predatory reptiles.

years, the land accumulated snow and ice, which had an additional cooling effect on the climate.

Most warmth-loving species, especially those living in the Tethys Sea, disappeared when the Cretaceous ended. The most temperature sensitive Tethyan fauna suffered the heaviest extinction rates. Species that were so successful in the warm waters of the Tethys dramatically declined when ocean temperatures dropped. Afterward, marine species acquired a more modern appearance as ocean bottom temperatures continued to plummet.

THE NEW RED SANDSTONE

The Triassic witnessed a complete retreat of marine waters from the land as the continents continued to rise. Abundant terrestrial redbeds and thick beds of gypsum and salt were deposited in the abandoned basins. Also, the amount of land covered with deserts was much greater in the Triassic than today, as indicated by a preponderance of red rocks composed of terrestrial sandstones and shales now exposed in the mountains and canyons in the western United States. Terrestrial redbeds covered a region from Nova Scotia to South Carolina and the Colorado Plateau (Fig. 10-11).

Figure 10-11   Chugwater redbeds on the east side of the Red Fork of the Powder River, Johnson County, Wyoming. Photo by N. H. Darton, courtesy of USGS

Redbeds are also common in Europe, where in northwestern England they are particularly well developed. Over northern and western Europe the terrestrial redbeds are characterized by a nearly fossil-free sequence called the New Red Sandstone, named for a sedimentary formation in Scotland famous for its dinosaur footprints. The unit shows a continuous gradation from Permian to Triassic in the region, with no clear demarcation between the two periods.

The wide occurrences of red sediments probably resulted from massive accumulations of iron supplied by one of the most intense intervals of igneous activity the world has ever known. Air trapped in ancient tree sap suggests a greater abundance of atmospheric oxygen, which oxidized the iron to form the mineral hematite, named so because of its blood-red color.

The mountain belts of the Cordilleran of North America, the Andean of South America, and the Tethyan of Africa-Eurasia contain thick marine deposits of Triassic age. The Cordilleran and Andean belts were created by the collision of east Pacific plates with the continental margins of the new American plates, formed when Pangaea rifted apart. The Tethyan belt formed when Africa collided with Eurasia, raising the Alps, which contain an abundantly fossil-bearing Triassic section. The Tethys Sea, located in the tropics during the Triassic, contained widespread coral reefs that uplifted to form the Dolomites and Alps during the collision of Africa and Eurasia in the Cenozoic.

Late in the Triassic, an inland sea began to flow into the west-central portions of North America. Accumulations of marine sediments eroded from the Cordilleran highlands to the west, often referred to as the ancestral Rockies, were deposited on the terrestrial redbeds of the Colorado Plateau. Important reserves of phosphate used for fertilizers were precipitated in the late Permian and early Triassic in Idaho and adjacent states. Huge sedimentary deposits of iron were also laid down. The ore-bearing rocks of the Clinton Iron Formation, the chief iron producer in the Appalachian region from Alabama to New York, were deposited during this time.

Evaporite accumulation peaked during the Triassic, when the supercontinent Pangaea was just beginning to rift apart. Evaporite deposits form when shallow brine pools generally replenished by seawater evaporate. Few evaporite deposits date beyond 800 million years ago, however, probably because most of the salt was buried or recycled into the sea. Ancient evaporite deposits exist as far north as the Arctic region, indicating that these areas were once closer to the equator or that the global climate was considerably warmer in the past.

TRIASSIC BASALTS

Over the past 250 million years, 11 episodes of massive flood basalt volcanism have occurred worldwide (Fig. 10-12 and Table 10-1). They were

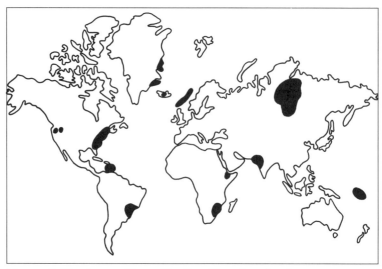

Figure 10-12 Areas affected by flood basalt volcanism.

relatively short-lived events, with major phases generally lasting less than 3 million years. These large eruptions created a series of overlapping lava flows, giving many exposures a terracelike appearance known as traps, from the Dutch word for "stairs."

Many flood basalts lie near continental margins, where great rifts began to separate the present continents from Pangaea near the end of the Triassic. These massive outpourings of basalt reflect one of the greatest crustal movements in the history of the planet. The continents probably traveled

TABLE 10–1 FLOOD BASALT VOLCANISM AND MASS EXTINCTION

Volcanic Episode	Million years ago	Extinction Event	Million years ago
Columbian River, U.S.A.	17	Low–mid Miocene	14
Ethiopian	35	Upper Eocene	36
Deccan, India	65	Maastrichtian	65
		Cenomanian	91
Rajmahal, India	110	Aptian	110
Southwest African	135	Tithonian	137
Antarctica	170	Bajocian	173
South African	190	Pliensbachian	191
E. North American	200	Rhaetian/Norian	211
Siberian	250	Guadalupian	249

much faster than they do today because of more vigorous plate motions, resulting in tremendous volcanic activity.

Triassic basalts common in eastern North America indicate the formation of a rift that separated the continent from Eurasia. The rift later breached and flooded with seawater, forming the infant North Atlantic Ocean. The Indian Ocean formed when a rift separated the Indian subcontinent from Gondwana. By the end of the Triassic, India drifted free and began its trek toward southern Asia. Meanwhile, Gondwana drifted northward, leaving Australia still attached to Antarctica, in the southern polar region.

Huge lava flows and granitic intrusions occurred in Siberia, and extensive lava flows covered South America, Africa, and Antarctica as well. Southern Brazil was paved with three-quarters of a million square miles of basalt, constituting the largest lava field in the world. Great floods of basalt, upwards of 2,000 feet or more thick, covered large parts of Brazil and Argentina when the South American plate overrode the Pacific plate, and the subduction fed magma chambers underlying active volcanoes. Basalt flows also blanketed a region from Alaska to California.

Near the end of the Triassic, North and South America began to move away from each other; India, nestled between Africa and Antarctica, began to separate from Gondwana; the Indochina block collided with China; and a great rift began to divide North America from Eurasia. The rifting of continents radically altered the climate and set the stage for the extraordinary warm periods that followed.

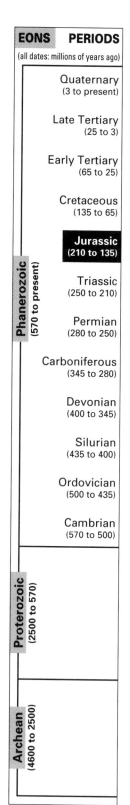

EONS
(all dates: millions of years ago)

Phanerozoic
(570 to present)

Proterozoic
(2500 to 570)

Archean
(4600 to 2500)

PERIODS

Quaternary
(3 to present)

Late Tertiary
(25 to 3)

Early Tertiary
(65 to 25)

Cretaceous
(135 to 65)

Jurassic
(210 to 135)

Triassic
(250 to 210)

Permian
(280 to 250)

Carboniferous
(345 to 280)

Devonian
(400 to 345)

Silurian
(435 to 400)

Ordovician
(500 to 435)

Cambrian
(570 to 500)

11

JURASSIC BIRDS

The Jurassic period, from 210 to 135 million years ago, was named for the limestones and chalks of the Jura Mountains in northwest Switzerland. Early in the period, Pangaea began rifting apart into the beginnings of the present continents, forming the Atlantic, Indian, and Arctic oceans. Mountains created by upheavals during previous periods were leveled by erosion, and inland seas invaded the continents, providing additional offshore habitats for a bewildering assortment of marine species (Fig. 11-1). By now, terrestrial faunas had attained the basic composition they would keep until the dinosaurs became extinct.

The dinosaurs were highly diversified by this time and reached their maximum size, becoming the largest terrestrial animals ever to live. There was a warm, moist climate, as suggested by widespread plant growth and coal formation. The beneficial climate and magnificent growing conditions contributed to the giant size of many dinosaur species, many of which became extinct at the end of the period. Reptiles were extremely successful and occupied land, sea, and air. Mammals were small, rodentlike creatures, sparsely populated and scarcely noticed. The first flight-worthy birds appeared and shared the skies with flying reptiles called pterosaurs.

Figure 11-1   Marine flora and fauna of the late Jurassic. Courtesy of Field Museum of Natural History, Chicago

THE EARLY BIRDS

Birds first appeared in the Jurassic about 150 million years ago, although some accounts push their origin as far back as the late Triassic, 225 million years ago. They descended from the thecodonts, the same ancestors of dinosaurs and crocodiles, and consequently birds are often referred to as "glorified reptiles." They were warm-blooded to obtain the maximum metabolic efficiency needed for sustained flight but retained the reptilian mode of reproduction by laying eggs. This ability to maintain a warm body temperature has led to speculation that some dinosaur species with similar skeletons were warm-blooded as well.

Archaeopteryx (Fig. 11-2), from the Greek for "ancient wing," was the earliest known fossil bird. It was about the size of a modern pigeon and appeared to be a transitional species between reptiles and true birds. Archaeopteryx was first thought to be a small dinosaur until fossils showing impressions of feathers were found in a unique limestone formation in Bavaria, Germany in 1863. The discovery sparked a long-standing contro-

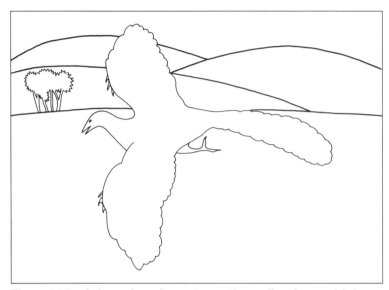

Figure 11-2   A Jurassic archaeopteryx, the earliest known bird.

versy. Prominent 19th-century geologists claimed Archaeopteryx was a hoax, and the feather impressions were simply etched into the rock containing the fossil. However, an Archaeopteryx fossil discovered in 1950 from the same Bavarian formation produced a well-preserved specimen that clearly showed feather impressions.

Although Archaeopteryx had many of the accouterments necessary for flight, it likely was a poor flyer and might have flown only short distances. It probably achieved flight by running along the ground with its wings outstretched and then glided for a brief moment or leapt from the ground while flapping its wings to catch an insect flying by. Archaeopteryx had teeth, claws, a long bony tail, and many of the skeletal features of a small dinosaur but lacked hollow bones for light weight. Its feathers were outgrowths of scales and probably originally functioned as insulation. Many bird species retained their teeth until the end of the Cretaceous.

Upon mastering the skill of flight, birds quickly radiated into all environments, and their superior adaptability enabled them to compete successfully with the pterosaurs, possibly leading to that reptile's extinction. Giant flightless land birds appeared early in the avian fossil record. Their wide distribution is further evidence for the existence of Pangaea, since these birds would then have had to walk from one corner of the world to another.

After being driven into the air by carnivoruos dinosaurs and kept there by hunting mammals, birds found life a lot easier on the ground once this threat was eliminated because they had to expend much energy to remain airborne. Some birds also successfully adapted to a life in the sea. Certain diving ducks are specially equipped for "flying" underwater to catch fish. Penguins, for example, are flightless birds that have taken to life in the water and are well adapted to survive in the Antarctic.

THE PTEROSAURS

Pterosaurs (Fig. 11-3), including the ferocious-looking pterodactyl, were flying reptiles with wingspans up to 40 feet and more. They originated in the early Jurassic and appear to have been the largest animals that ever flew, dominating the skies for more than 120 million years. Their wings were constructed by elongating the fourth finger of each forelimb, which supported the front edge of a membrane that stretched from the flank of the body to the fingertip, leaving the other fingers free for such purposes as climbing trees.

By comparison, a bat's wing is constructed by lengthening and splaying all fingers and covering them with membrane. The wing membranes might have originally served as a cooling mechanism used to regulate body temperature by fanning the forelimbs. Why pterosaurs took to the air in the first place is still a mystery. Their ancestors might have grown skin flaps for jumping out of trees like flying squirrels.

The pterosaurs resembled both birds and bats in their overall structure and proportions, with the smallest species roughly the size of a sparrow. Like birds, they had hollow bones to conserve weight for flight. The larger

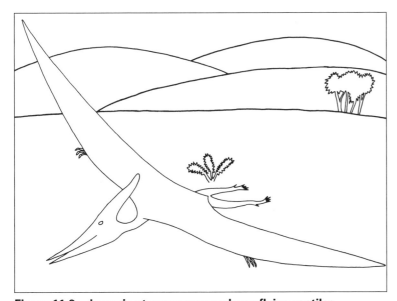

Figure 11-3   Jurassic pterosaurs were large flying reptiles.

pterosaurs were proportioned similar to a modern hang-glider and weighed about as much as the human pilot. Many pterosaurs had tall crests on their skulls, which possibly functioned as a forward rudder to steer them in flight.

Pterosaurs might have achieved flight by jumping off cliffs and riding the updrafts, by climbing trees and diving into the wind, or by gliding across the tops of wave crests like modern albatrosses. They could have trotted along the ground flapping their wings and taking off gooney bird-fashion

TABLE 11–1   CONTINENTAL DRIFT

Geologic division (millions of years ago)		Gondwana	Laurasia
Quaternary	3		Opening of Gulf of California
Pliocene	11	Spreading begins near Galapagos Islands	Spreading directions change in eastern Pacific
		Opening of the Gulf of Aden	
			Birth of Iceland
Miocene	26		
		Opening of Red Sea	
Oligocene	37		
		Collision of India with Eurasia	Spreading begins in Arctic Basin
Eocene	54		Separation of Greenland from Norway
		Separation of Australia from Antarctica	
Paleocene	65	Separation of New Zealand from Antarctica	Opening of the Labrador Sea
		Separation of Africa from Madagascar and South America	Opening of the Bay of Biscay
			Major rifting of North America from Eurasia
Cretaceous	135		
		Separation of Africa from India, Australia, New Zealand, and Antarctica	
Jurassic	180		
			Separation of North America from Africa begins
Triassic	250		

or simply stood on their hind legs, caught a strong breeze, and with a single flap of their huge wings and a kick of their powerful legs become airborne. The pterosaur probably spent most of its time aloft riding air currents like present-day condors.

When landing, it simply stalled near the ground, gently touching down on its hind legs like a hang-glider does. While on the ground, pterosaurs might have been ungainly walkers, sprawling about on all fours like a bat. However, fossil pterosaur pelvises seem to indicate that the hind legs extended straight down from the body, enabling the reptiles to walk upright on two feet. They could then trot along for short bursts to gather speed for takeoff. That the animals could actually fly is not doubted, and they went on to become the greatest animal aviators the world has ever known.

THE GIANT DINOSAURS

The oldest dinosaurs originated on the southern continent Gondwana when the last glaciers from the great Permian ice age were departing. They ventured to all major continents, and their distribution throughout the world is strong evidence for continental drift (Table 11-1). At the time the dinosaurs came into existence all continents were assembled into Pangaea. Early in the Jurassic, it began to rift apart, and the continents drifted toward their present locations. Except for a few temporary land bridges, the oceans that filled the rifts between the newly formed continents provided a barrier to further dinosaur migration. At this time, almost identical species lived in North America, Europe, and Africa (Fig. 11-4).

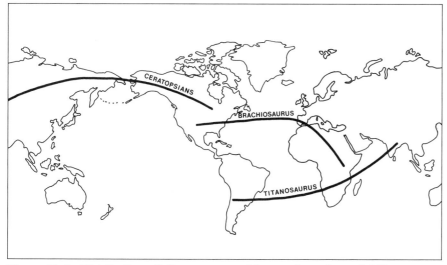

Figure 11-4   Distribution of dinosaurs as evidence of continental drift.

Figure 11-5   Restoration of dinosaur bones at Dinosaur National Monument, Utah. Courtesy of National Park Service

The success of the dinosaurs is exemplified by their extensive range, wherein they occupied a wide variety of habitats and dominated all other forms of land-dwelling animals. Indeed, if the dinosaurs had not become extinct, mammals would never have achieved dominance over the Earth and humans would not have come into existence. The dinosaurs would have continued to suppress further advancement of the mammals, which would have remained small, nocturnal creatures, keeping out from underfoot of the dinosaurs.

Over 500 dinosaur genera have been discovered over the last 175 years. Among the largest dinosaur species were the brachiosaurs and the apathosaurs, formerly called brontosaurs. They were sauropods with long, slender tails and necks, and the front legs were longer than the hind legs. Their fossils are found in Colorado and Utah, southwestern Europe, and East Africa, some species probably having traveled to Africa by way of Europe.

The Jurassic Morrison Formation, a famous bed of sediments in the Colorado Plateau region, has yielded some of the largest dinosaur fossils. Many of the best specimens are displayed at Dinosaur National Monument near Vernal, Utah (Fig. 11-5). Perhaps the tallest and heaviest dinosaur ever discovered is the 80-ton ultrasaurus, which could look down onto the roof of a five-story building. Seismosaurus, meaning "earth-shaker," was the longest known dinosaur, possibly reaching a length of more than 140 feet from its head, which was supported by a long, slender neck, to the tip of its even longer, whiplike tail.

Dinosaurs attained their largest sizes and longest life spans during the Jurassic. Large reptiles possess the power of almost unlimited growth. Adults never cease growing entirely but continue to increase in size until disease or accident take their lives. The giant Komodo dragon lizards of southeast Asia, for example, grow to over 300 pounds and prey on monkeys, pigs, and deer. Reptiles with their continuous growth achieved a measure of eternal youth, whereas mammals grow rapidly to adulthood and then slowly degenerate and die.

A large body allows a cold-blooded reptile to retain its body temperature for long periods. A large body retards heat loss better than a small one because it has a better surface-area-to-volume ratio. Thus, the animal is less susceptible to short-term temperature variations such as cool nights or cloudy days. Conversely, a large reptile takes much longer to warm up from an extended cold period than a small one. Muscles also generate body heat, although for reptiles it is only about a quarter of that produced by mammals during exertion. A high steady body temperature maintains an efficient metabolism, and higher temperatures enhance the output of muscles.

Figure 11-6   Tyrannosaurus rex was the greatest land carnivore that ever lived.

Therefore, the performance of some large dinosaurs probably could match that of large mammals.

The generally warm climate of the Mesozoic produced excellent growing conditions for lush vegetation, including ferns and cycads, to satisfy the diets of the plant-eating dinosaurs. The herbivorous dinosaurs developed a large stomach to digest the tough, fibrous fronds, requiring an enormous body to carry it around. The dinosaurs grew to such giants probably for the same reasons that large ungulates like the rhinoceros and elephant are so big. Most large dinosaurs were herbivores that consumed huge quantities of coarse cellulose that required much time to digest. This required a large stomach and therefore a large body to carry it about.

Some dinosaur species swallowed cobbles called gizzard stones, similar to the grit used by many modern birds, to pulp the coarse vegetation in their stomachs. The digestive juices further broke down the rough material, and this long fermentation process required a large storage capacity. The rounded polished stones were often left in a heap where the dinosaur died. Sometimes deposits of these stones lie atop exposed Mesozoic sediments.

The large size of the herbivores spurred the evolution of giant carnivorous dinosaurs to prey on them, such as tyrannosaurus rex (Fig. 11-6), perhaps the fiercest carnivore of them all. The giant dinosaurs were prevented from

Figure 11-7   Allosaurs were among several dinosaur genera that went extinct at the end of the Jurassic. Courtesy of National Museums of Canada

growing any larger due to the force of gravity. When an animal doubles its size, the weight on its bones is four times greater. The only exceptions were dinosaurs living permanently in the sea. As with modern whales, some of which are even larger than the biggest dinosaurs, the buoyancy of seawater kept the weight off their bones. If an animal accidentally beached itself, as whales sometimes do, it suffocated because its bones were unable to support the weight of the body, crushing the lungs.

Many families of large dinosaurs, including apathosaurs, stegosaurs, and allosaurs (Fig. 11-7), became extinct at the end of the Jurassic. Following the extinction, the population of small animals exploded, as species occupied niches vacated by the large dinosaurs. Most of the surviving species were aquatic, confined to freshwater lakes and marshes, and small land-dwelling animals. Many of the small nondinosaur species were the same types that survived the dinosaur extinction at the end of the Cretaceous, probably due to their large populations and ability to find places to hide.

THE BREAKUP OF PANGAEA

Throughout the Earth's history, continents appear to have undergone cycles of collision and rifting. Smaller continental blocks collided and merged into larger continents. Millions of years later, the continents rifted apart, and the chasms filled with seawater to form new oceans. The regions presently bordering the Pacific Basin apparently have never collided with each other. The Pacific Ocean is a remnant of an ancient sea called the Panthalassa, which narrowed and widened in response to continental breakup, dispersal, and reconvergence in the area occupied by the present Atlantic Ocean.

Several oceans have repeatedly opened and closed in the vicinity of the Atlantic Basin, while a single ocean has existed continuously in the area of the Pacific Basin. The Pacific plate was hardly larger than the United States after the breakup of Pangaea in the early Jurassic about 180 million years ago. The rest of the ocean floor consisted of other unknown plates that disappeared as the Pacific plate grew, and consequently no oceanic crust is older than Jurassic in age.

Early in the Jurassic, North America separated from South America, and a rift divided the North American and Eurasian continents. India, nestled between Africa and Antarctica, drifted away from Gondwana, and Antarctica—still attached to Australia—swung away from Africa toward the southeast, forming the proto-Indian Ocean. The rift separating the continents breached and flooded with seawater, forming the infant North Atlantic Ocean. Many ridges of the Atlantic's spreading seafloor remained above sea level, creating a series of stepping-stones for the migration of animals between the Old and New Worlds.

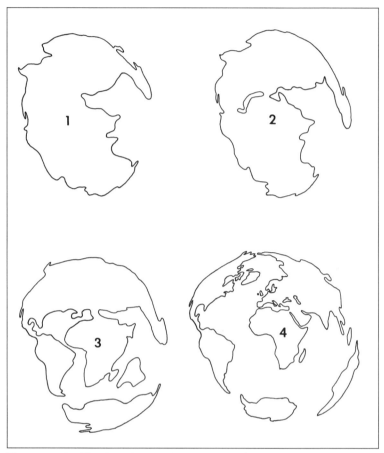

Figure 11-8 **The breakup and dispersal of the continents: 1) Pangaea, 200 million years ago; 2) opening of the North Atlantic, 160 million years ago; 3) the continents at the height of the dinosaur age, 80 million years ago; and 4) the present locations of the continents.**

When Pangaea began to separate into today's continents (Fig. 11-8), a great rift developed in the present Caribbean. It sliced northward through the continental crust connecting North America, northwest Africa, and Eurasia and began to open the Atlantic Ocean. The process took several million years along a zone hundreds of miles wide. The breakup of North America and Eurasia might have resulted from upwelling basaltic magma that weakened the continental crust. Many flood basalts exist near continental margins, evidence of where rifts separated the present continents. The episodes of flood basalt volcanism were short-lived events, with major phases generally lasting less than 3 million years.

About 125 million years ago, the infant North Atlantic obtained a depth of about 2.5 miles and was bisected by an active midocean ridge system

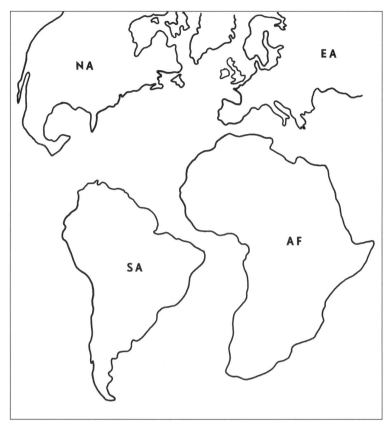

**Figure 11-9   The opening of the Atlantic Ocean by the middle Creta-
ceous, showing North America (NA), South America (SA), Europe
(EA), and Africa (AF).**

producing new oceanic crust. At about the same time, the South Atlantic
began to form, opening up like a zipper from south to north. The rift
propagated northward several inches per year, comparable to the plate
separation rate. The entire process of opening the South Atlantic was
completed in only about 5 million years. By 80 million years ago, the North
Atlantic had become a fully developed ocean (Fig. 11-9). Some 20 million
years later, the Mid-Atlantic rift progressed into the Arctic Basin, detaching
Greenland from Europe.

   After breakup, the continents traveled in spurts rather than drifting apart at
a constant speed. The rate of seafloor spreading in the Atlantic was matched
by plate subduction in the Pacific, where one plate dives under another,
forming a deep trench (Table 11-2). This is why the oceanic crust of the Pacific
Basin dates back no farther than the early Jurassic. A high degree of geologic
activity around the Pacific rim produced practically all the mountain ranges
facing the Pacific and the island arcs along its perimeter.

TABLE 11–2   THE WORLD'S OCEAN TRENCHES

Trench	Depth (miles)	Width (miles)	Length (miles)
Peru/Chile	5.0	62	3700
Java	4.7	50	2800
Aleutian	4.8	31	2300
Middle America	4.2	25	1700
Marianas	6.8	43	1600
Kuril/Kamchatka	6.5	74	1400
Puerto Rico	5.2	74	960
South Sandwich	5.2	56	900
Philippine	6.5	37	870
Tonga	6.7	34	870
Japan	5.2	62	500

Much of western North America was assembled from island arcs and other crustal debris skimmed off the Pacific plate as the North American plate continued heading westward. Northern California is a jumble of crustal fragments assembled within about 200 million years ago. A nearly complete slice of ocean crust, the type that is shoved up on the continents

Figure 11-10   A plunging syncline in the Bright Dot Formation, Sierra Nevada Range, Mono County, California.   Photo by C. D. Rinehart, courtesy of USGS

by drifting plates, sits in the middle of Wyoming. The Nevadan orogeny produced the Sierra Nevada Range in California (Fig. 11-10) from the middle to late Jurassic.

The breakup of Pangaea compressed the ocean basins, causing a rise in sea level and a transgression of the seas onto the land. In addition, an increase in volcanism flooded the continental crust with vast amounts of basalt. The rise in volcanic activity also increased the carbon dioxide content of the atmosphere, resulting in a strong greenhouse effect that led to the warm Mesozoic climate.

Continental breakup and dispersal might also have contributed to the extinction of many dinosaur species. The shifting of continents changed global climate patterns and brought unstable weather conditions to many parts of the world. Massive lava flows from perhaps the most volcanically active period since the early days of the Earth might have dealt a major blow to the climatic and ecological stability of the planet.

SEA

Figure 11-11   The middle Jurassic inland sea in North America.

MARINE TRANSGRESSION

Throughout most of the Earth's history, several crustal plates constantly in motion reshaped and rearranged continents and ocean basins. When continents broke up, they overrode ocean basins, which compressed the seas, thereby raising global sea levels several hundred feet. The rising seas inundated low-lying areas inland of the continents, dramatically increasing the shoreline and shallow-water marine habitat area, which in turn supported many more species.

The vast majority of marine species live on continental shelves, shallow-water portions of islands, and subsurface rises generally less than 600 feet deep. The richest shallow-water faunas live in the tropics, which contain large numbers of highly specialized organisms. Species diversity also depends on the shapes of the continents, the width of shallow continental margins, the extent of inland seas, and the presence of coastal mountains, all of which are affected by continental motions.

Extensive mountain building is also associated with the movement of crustal plates. The upward thrust of continental rocks alters patterns of river

Figure 11-12 The Jurassic Morrison Formation in the Uinta Mountains, Utah, which holds abundant fossil dinosaur bones.

drainage and climate, which in turn affects terrestrial habitats. The raising of land to higher elevations, where the air is thin and cold, spurs the growth of glacial ice, especially in the higher latitudes. Furthermore, continents scattered in all parts of the world interfere with ocean currents, which distribute heat over the globe.

During the Jurassic and continuing into the Cretaceous, an interior sea flowed into the west-central portions of North America (Fig. 11-11). Massive accumulations of marine sediments eroded from the Cordilleran highlands to the west and were deposited on the terrestrial redbeds of the Colorado Plateau, forming the Jurassic Morrison Formation, well-known for fossil bones of large dinosaurs (Fig. 11-12). Eastern Mexico, southern Texas, and Louisiana were also flooded. Seas invaded South America, Africa, and Australia as well.

The continents were flatter, mountain ranges were lower, and sea levels were higher. Thick deposits of sediment that filled the seas flooding North America were uplifted and eroded, giving the western United States its impressive scenery. Reef building was intense in the Tethys Sea, and thick deposits of limestone and dolomite were laid down in the interior seas of Europe and Asia, later to be uplifted during one of geologic history's greatest mountain building episodes.

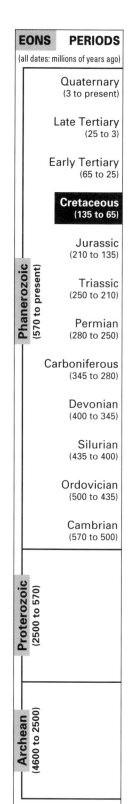

12

CRETACEOUS CORALS

The Cretaceous, which ran from 135 to 65 million years ago, was named from the Latin word *creta,* meaning chalk, due to the vast deposits of carbonate rock laid down worldwide at this time. It was the warmest period of the Phanerozoic as evidenced by extensive coral reefs, which built massive limestone deposits (Fig. 12-1). Coral reefs and other tropical biota, for which bright sunlight and warm seas are essential, ranged far into higher latitudes. They fringed the continents and covered the tops of extinct marine volcanoes.

The warm climate was particularly advantageous to the ammonites (coiled-shell cephalopods), which grew to tremendous size, becoming the predominant creatures of the Cretaceous seas. The dinosaurs did exceptionally well during the Cretaceous, but along with the ammonites and many other species, they mysteriously vanished at the end of the period. The extinction was apparently caused by a cataclysm that created intolerable living conditions for most species on Earth.

THE AGE OF AMMONITES

Coral reefs were the most widespread during the Cretaceous, ranging a thousand miles farther away from the equator, whereas today they are restricted to the tropics (Fig. 12-2). The corals began constructing reefs in

Figure 12-1 Limestones of the Hawthorn and Ocala formations, Marion County, Florida. Photo by G. H. Espenshade, courtesy of USGS

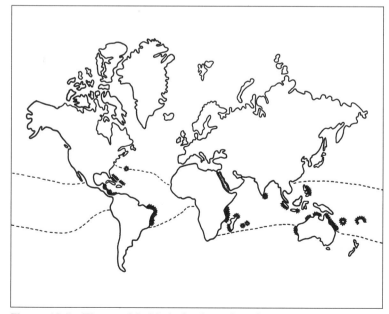

Figure 12-2 The worldwide belt of coral reefs.

the early Paleozoic. The hexacorals from the Triassic to the present were the major reef builders of the Mesozoic and Cenozoic seas. The corals constructed barrier reefs and atolls, which were massive structures composed of calcium carbonate lithified into limestone. The Great Barrier Reef, stretching more than 1,200 miles along the northeast coast of Australia, is the largest feature built by living organisms.

The cephalopods were the most spectacular, diversified, and successful marine invertebrates of the Mesozoic seas. The nautiloids grew to lengths of 30 feet or more, and with straight, streamlined shells they were among the swiftest creatures of the deep. The ammonites, the most significant cephalopods, had a variety of coiled-shell forms identified by their complex suture patterns, making them the most important guide fossils for dating Mesozoic rocks.

Unfortunately, after surviving the critical transition from the Permian to the Triassic and recovering from serious setbacks during the Mesozoic, the ammonites suffered final extinction at the end of the Cretaceous, when the recession of the seas reduced their shallow-water habitats worldwide. The ammonites declined over a period of about 2 million years, possibly becoming extinct 100,000 years prior to the end of the Cretaceous.

A fast-swimming, shell-crushing marine predator called ichthyosaur, Greek for "fish lizard," apparently preyed on ammonites by first puncturing the shell from the ammonite's blind side, causing it to fill with water and sink to the bottom, where the attack could then be made head on. These

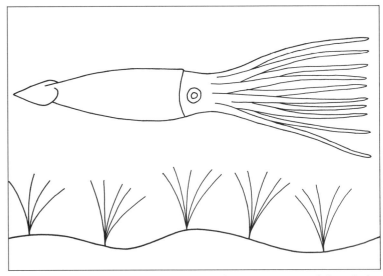

Figure 12-3   The squids were among the most successful cephalopods.

highly aggressive predators might have caused the extinction of most ammonite species before the Cretaceous was out.

All shelled cephalopods were absent in the Cenozoic seas except the nautilus, found exclusively in the deep waters of the Indian Ocean and the ammonite's only living relative, along with shell-less species, including cuttlefish, octopus, and squids (Fig. 12-3). The squids competed directly with fish, which were little affected by the extinction. Other major marine groups that disappeared at the end of the Cretaceous include the rudists, which were huge coral-shaped clams, and other types of clams and oysters. The gastropods, including snails and slugs, increased in number and variety throughout the Cenozoic and presently are second only to insects in diversity.

THE ANGIOSPERMS

The Mesozoic was a time of transition, especially for plants, which showed little resemblance at the beginning of the era to those at the end, when they more closely resembled present-day vegetation. The gymnosperms originated in the Permian and bore seeds without fruit coverings, including conifers, ginkgoes, and palmlike cycads. The true ferns prospered in the higher latitudes, whereas today they live only in the warm tropics.

The cycads, which resembled palm trees, were also highly successful and ranged across all major continents, possibly contributing to the diets of the plant-eating dinosaurs. The ginkgo, of which the maidenhair tree in eastern China is the only living relative, might have been the oldest genus of the seed plants. Also dominating the landscape were conifers up to 5 feet across and 100 feet long. Their petrified trunks are especially plentiful at Yellowstone National Park (Fig. 12-4).

About 110 million years ago, vegetation in the early Cretaceous underwent a radical change with the introduction of the angiosperms, flowering plants that evolved alongside pollinating insects. They might have originally exploited the weedy rift valleys that formed as Pangaea split apart. The earliest angiosperms appear to have been large plants, growing as tall as magnolia trees. However, fossils discovered in Australia suggest that the first angiosperms there and perhaps elsewhere were small herb-like plants. Within a few million years after their introduction, the efficient flowering plants crowded out the once abundant ferns and gymnosperms.

The angiosperms were distributed worldwide by the end of the Cretaceous, and today they include about a quarter-million species of trees, shrubs, grasses, and herbs. The plants offered pollinators, such as honey bees and birds, brightly colored and scented flowers, and sweet nectar. The unwary intruder was dusted with pollen, which it transported to the next flower it visited for pollination. Many angiosperms also depended on

Figure 12-4 Petrified tree stumps in Yellowstone National Park, Wyoming.
Courtesy of National Park Service

animals to spread their seeds, which were encased in tasty fruit that passed through the body and dropped some distance away.

Near the end of the Cretaceous, forests extended into the polar regions far beyond the present tree line. The most remarkable example is a well preserved fossil forest on Alexander Island, Antarctica. To survive the harsh arctic conditions, trees had to develop a means of protection against the cold, since plants are more sensitive to the lack of heat than the absence of sunlight. They probably adapted mechanisms for intercepting the maximum amount of sunlight during a period when global temperatures were considerably warmer than today.

The cone-bearing plants prominent during the entire Mesozoic occupied only a secondary role during the Cenozoic. Tropical vegetation that was widespread during the Mesozoic withdrew to narrow regions around the equator in response to a colder, drier climate, a result of a general uplift of the continents and the draining of the interior seas. Forests of giant hardwood trees that grew as far north as Montana were replaced by scraggly conifers, a further indication of a cooler climate.

The rise of the angiosperms near the end of the Cretaceous might even have contributed to the death of the dinosaurs and certain marine species.

By absorbing large quantities of carbon dioxide from the atmosphere, they caused a drop in global temperatures. Also contributing to the dinosaur's downfall, forests of broadleaf trees and shrubs that were a favorite food of the dinosaurs apparently disappeared just prior to the ending of the Cretaceous.

THE LARAMIDE OROGENY

Beginning about 80 million years ago, a large part of western North America uplifted, and the entire Rocky Mountain region from northern Mexico into Canada rose nearly a mile above sea level (Fig. 12-5). This mountain-building episode, called the Laramide orogeny, resulted from the subduction of oceanic crust beneath the West Coast of North America, causing an increase in crustal buoyancy. The Canadian Rockies consist of slices of sedimentary rock that were successively detached from the underlying basement rock and thrust eastward on top of each other. A region between the Sierra Nevada and the southern Rockies experienced a spurt of uplift over the past 20 million years, raising the area over 3,000 feet.

During the late Cambrian, the future Rocky Mountain region was near sea level. Farther west, within about 400 miles of the coast, a mountain belt comparable to the present Andes formed above a subduction zone during

Figure 12-5 Lake Sherburne Valley with the glaciated Rocky Mountains in the background, Glacier National Park, Montana. Photo by H. E. Malde, courtesy of USGS

Figure 12-6   Death Valley showing salt pan, alluvial fans, and fault scarp, Inyo County, California.
Photo by H. Drews, courtesy of USGS

the 80 million years prior to the Laramide. It might have been responsible for the Cretaceous Sevier orogeny that created the Overthrust Belt in Utah and Nevada. A region from eastern Utah to the Texas panhandle that deformed during the late Paleozoic, Ancestral Rockies orogeny was completely eroded by the time of the Laramide. The Rocky Mountain foreland region subsided as much as 2 miles between 85 million and 65 million years ago and then rose well above sea level, reaching its present elevation around 30 million years ago.

To the west of the Rockies, numerous parallel faults sliced through the Basin and Range Province between the Sierra Nevada of California and the Wasatch Mountains of Utah, resulting in a series of about 20 north-south-trending fault-block mountain ranges. The Basin and Range covers southern Oregon, Nevada, western Utah, southeastern California, and southern Arizona and New Mexico. The crust bounded by faults is literally broken into hundreds of steeply tilted blocks and raised nearly a mile above the basin, forming nearly parallel mountain ranges up to 50 miles long.

Death Valley (Fig. 12-6), which is presently 280 feet below sea level, the lowest place on the North American continent, sat several thousand feet higher during the Cretaceous. The region collapsed when the continental crust thinned from extensive block faulting, with one block of crust lying below another. The Great Basin area is a remnant of a broad belt of mountains and high plateaus that subsequently collapsed after the crust was pulled apart following the Laramide.

The rising Wasatch Range of north-central Utah and south Idaho is an excellent example of a north-trending series of faults, one below the other. The fault blocks extend for 80 miles, with a probable net slip along the west side of 18,000 feet. The Tetons of western Wyoming were upfaulted along the eastern flank and downfaulted to the west. The rest of the Rocky Mountains evolved by a process of upthrusting similar to the plate collision and subduction that raised the Andes Mountains of Central and South America. The Andes continue to rise due to an increase in crustal buoyancy caused by the subduction of the Nazca plate to the west beneath the South American plate.

CRETACEOUS WARMING

During the Cretaceous, plants and animals were especially prolific and ranged practically from pole to pole. The deep ocean waters, which are now near freezing, were about 15 degrees Celsius (60 degrees Fahrenheit) during the Cretaceous. The average global surface temperature was 10 to 15 degrees warmer than at present. Temperatures were also much warmer in the polar regions, with a temperature difference between the poles and the equator of only 20 degrees Celsius, or about half that of today.

The drifting of continents into warmer equatorial waters might have accounted for much of the mild climate during the Cretaceous. By the time of the initial breakup of the continents about 180 million years ago, the climate had begun to warm dramatically. The continents were flatter, with lower mountains and higher sea levels. Although the geography during this time was important, it did not account for all of the warming.

The movement of the continents was more rapid than today, with perhaps the most vigorous plate tectonics the world has ever known. About 120 million years ago, an extraordinary burst of submarine volcanism struck the Pacific Basin, releasing vast amounts of gas-laden lava onto the ocean floor. These volcanic spasms are evidenced by a collection of massive undersea lava plateaus that formed almost simultaneously, the largest of which, the Ontong Java, is about two-thirds the size of Australia. It contains at least 9 million cubic miles of basalt, enough to bury the United States under 3 miles of lava.

The surge of volcanism increased the production of oceanic crust as much as 50 percent. This rise in volcanic activity provided perhaps the greatest contribution to the warming of the Earth, producing 4 to 8 times the present amount of atmospheric carbon dioxide, and worldwide temperatures averaged 7.5 to 12.5 degrees Celsius higher than today.

TABLE 12–1   COMPARISON OF MAGNETIC REVERSALS WITH OTHER PHENOMENA (dates in millions of years)

Magnetic Reversal	Unusually Cold	Meteorite Activity	Sea Level Drops	Mass Extinctions
0.7	0.7	0.7		
1.9	1.9	1.9		
2.0	2.0			
10				11
40			37–20	37
70			70–60	65
130			132–125	137
160			165–140	173

For the next 40 million years, the Earth's geomagnetic field, which normally reverses polarity quite often on a geologic time scale (Table 12-1), stabilized and assumed a constant orientation due to several mantle plumes that produced tremendous basaltic eruptions. This greater volcanic activity increased the carbon dioxide content of the atmosphere, producing the warmest global climate in 500 million years. Carbon dioxide also provided an abundant source of carbon for green vegetation and contributed to its prodigious growth, supplying substantial diet for herbivorous dinosaurs.

Polar forests extended into latitudes 85 degrees north and south of the equator, as indicated by fossilized remains of an ancient forest that thrived on the now frozen continent of Antarctica. Evidence of a warm climate that supported lush vegetation is provided by coal seams that run through the Transantarctic Mountains that are among the most extensive coal beds in the world. Alligators and crocodiles lived in the high latitudes as far north as Labrador, whereas today they are confined to warm tropical areas. The duck-billed hadrasaurs also lived in the Arctic regions of the Northern and Southern Hemispheres.

The positions of the continents might have contributed to the warming of the climate during most of the Mesozoic. Continents bunched together near the equator during the Cretaceous allowed warm ocean currents to carry heat poleward. High-latitude oceans are less reflective than land and absorb more heat, further moderating the climate.

THE INLAND SEAS

In the late Cretaceous and early Tertiary, land areas were inundated by high sea levels that flooded continental margins and formed great inland seas,

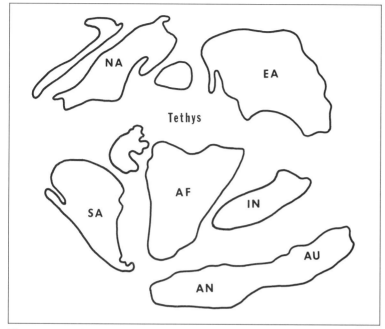

Figure 12-7 Distribution of continents around the Tethys Sea during the late Cretaceous, showing North America (NA), South America (SA), Europe (EA), Africa (AF), India (IN), Antarctica (AN), and Australia (AU).

some of which split continents in two. Seas divided North America in the Rocky Mountain and high plains regions, South America was cut in two in the region that later became the Amazon basin, and Eurasia was split by the joining of the Tethys Sea and the newly-formed Arctic Ocean.

The oceans of the Cretaceous were also interconnected in the equatorial regions by the Tethys and Central American seaways (Fig. 12-7), providing a unique circumglobal oceanic current system that made the climate equable. Mountains were lower and sea levels higher, and the total land surface declined to perhaps half its present size. The Appalachians, which were an imposing mountain range at the beginning of the Triassic, were eroded down to stumps in the Cretaceous. Erosion toppled the once towering mountain ranges of Eurasia as well.

Great deposits of limestone and chalk were laid down in Europe and Asia, which is how the Cretaceous received its name. Seas invaded Asia, Africa, Australia, South America, and the interior of North America. About 80 million years ago, the Western Interior Cretaceous Seaway (Fig. 12-8) was a shallow body of water that divided the North American continent into the western highlands, comprising the newly forming Rocky Mountains

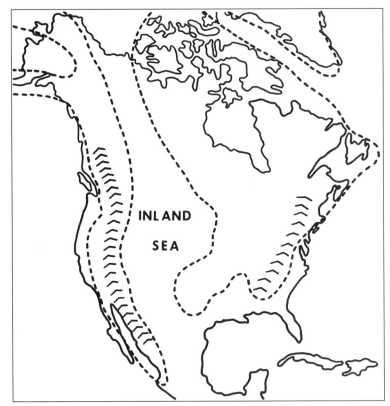

Figure 12-8 The Cretaceous interior sea of North America, where thick deposits of sediments were laid down.

Figure 12-9 Cretaceous Mancos shale and Mesaverde formations, Mesa Verde National Park, Colorado. Photo by L. C. Huff, courtesy of USGS

and isolated volcanoes, and the eastern uplands, consisting of the Appalachian Mountains.

Eastward of the rising Rocky Mountains was a broad coastal plain composed of thick layers of sediments eroded from the mountainous regions and extended to the western shore of the interior seaway. These sediment layers were later lithified and upraised and today are exposed as impressive cliffs in the western United States (Fig. 12-9). Along the coast and extending some distance inland were extensive wetlands, where dense vegetation grew in the subtropical climate. Inhabiting these areas were fishes, amphibians, aquatic turtles, crocodiles, and primitive mammals. The dinosaurs included herbivorous hadrasaurs and triceratops, and the carnosaurs that preyed on them.

Toward the end of the Cretaceous, North America and Europe were no longer in contact, except for a land bridge that spanned Greenland to the

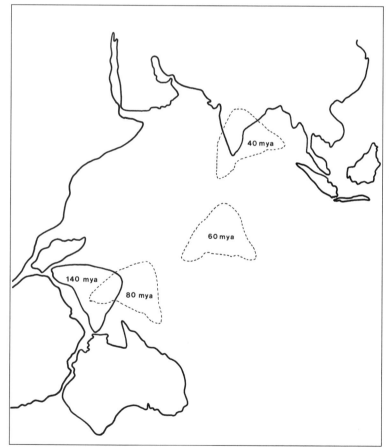

Figure 12-10 The drift of India, which collided with Asia about 45 million years ago.

north. The strait between Alaska and Asia narrowed, creating the practically landlocked Arctic Ocean. The South Atlantic continued to widen, with South America and Africa separated by over 1,500 miles of ocean. Africa moved northward, leaving behind Antarctica, which was still joined to Australia, and began to close the Tethys Sea.

Meanwhile, the northward-drifting subcontinent of India, traveling about 2 inches per year, narrowed the gap between itself and southern Asia (Fig. 12-10). During its journey after breakup with Gondwana, no known mammals appear to have existed in India until after the collision with Eurasia. Apparently, during the middle Cretaceous, Australia—still attached to Antarctica—wandered near the Antarctic Circle and acquired a thick mantle of ice. As Antarctica and Australia continued to move eastward, a rift developed that eventually separated them. Australia moved into the lower latitudes, while Antarctica drifted into the south polar region, accumulating a massive ice sheet.

When the Cretaceous ended, the seas regressed from the land because of lowered sea levels, and the climate grew colder. The last stage of the Cretaceous, called the Maestrichtian, was the coldest interval of the period. The decreasing global temperatures and increasing seasonal variation in the weather made the world stormier, with powerful gusty winds that wreaked havoc over the Earth.

There is no clear evidence of significant glaciation during this time. However, most warmth-loving species, especially those living in the Tethys Sea, disappeared when the Cretaceous came to an end. The extinctions appear to have been gradual, occurring over a period of 1 to 2 million years. Moreover, those species already in decline, including the dinosaurs and pterosaurs, might have been dealt a fatal blow from above.

AN ASTEROID IMPACT

One theory that attempts to explain the extinction of the dinosaurs and over 70 percent of other species at the end of the Cretaceous suggests that one or more large asteroids or comets struck the Earth with an equivalent explosive force of 100 trillion tons of dynamite or about a million eruptions of Mount St. Helens. The impact would have sent 500 billion tons of debris into the atmosphere and plunged the planet into environmental chaos. Generally, no animal heavier than 50 pounds survived the extinction, and a large body size appears to have been a severe disadvantage among terrestrial animals.

Following the impact, glowing bits of impact debris flying through the atmosphere set ablaze globe-wide forest fires, burning perhaps a quarter of all vegetation on the continents and turning a large part of the Earth into a smoldering cinder. A heavy blanket of dust and soot encircled the entire globe and lingered for months, cooling the planet and halting photosynthesis.

A catastrophe on this scale would have destroyed most terrestrial habitats and caused extinctions of tragic proportions. Species living in the tropics that relied on steady warmth and sunshine, like the coral reef communities, were especially hard hit. For example, the rudists, which built reef-like structures, completely died out, along with half of all bivalve genera.

A massive bombardment of meteorites also might have stripped away the upper atmospheric ozone layer, bathing the Earth in the sun's deadly ultraviolet rays. The increased radiation would have killed land plants and animals as well as primary producers in the surface waters of the ocean. The mammals, which were no larger than rodents, coexisted with the dinosaurs for more than 100 million years. But because they were mostly nocturnal and remained in their underground burrows in the daylight hours, only coming out at night to feed, the mammals would have been spared the onslaught of ultraviolet radiation during the daytime.

In the aftermath of the bombardment, the Earth would have succumbed to a year of darkness, under a thick brown smog of nitrogen oxide. Surface waters, poisoned by trace metals leached from the soil and rock, and global rains as corrosive as battery acid would have destroyed terrestrial life forms. Plants that survived as seeds and roots would have been relatively unscathed. The high acidity levels would have dissolved the calcium carbonate shells of marine organisms, while those with silica shells would have survived as they have done during other crises. Land animals living in burrows would have been well protected, and creatures living in lakes buffered against the acid would have survived the meteorite impacts quite well.

The impacts also could have caused widespread extinctions of microscopic marine plants called calcareous nannoplankton, which produce a sulfur compound that aids in cloud formation. With the death of these creatures, cloud cover would have decreased dramatically, triggering a global heat wave extreme enough to kill off the dinosaurs and most marine species. This contention is supported by the fossil record, which shows that ocean temperatures rose 5 to 10 degrees Celsius for tens of thousands of years beyond the end of the Cretaceous. During this time, over a period of almost almost half a million years, more than 90 percent of the calcareous nannoplankton disappeared along with most marine life in the upper portions of the ocean.

Sixty-five-million-year-old sediments found at the boundary between the Cretaceous and Tertiary periods throughout the world (Fig. 12-11) contain shocked quartz grains with distinctive lamellae, common soot from global forest fires, rare amino acids known to exist only on meteorites, the mineral stishovite, a dense form of silica found nowhere except at known impact sites, and unique concentrations of iridium, a rare isotope of platinum

Figure 12-11 The Cretaceous-Tertiary rocks shown at the dark streak just above the white sandstone in the center near Rock Springs, Wyoming. Photo by R. W. Brown, courtesy of USGS

relatively abundant on meteorites and comets but practically nonexistent in the Earth's crust.

The geologic record holds other iridium anomalies thought to be associated with other giant meteorite impacts that also coincide with extinction episodes. However, they are not nearly as intense as the iridium concentrations at the end of the Cretaceous, which are as high as a thousand times normal background levels, suggesting that the end of Cretaceous extinction might have been a unique event in the history of life on Earth.

13

TERTIARY MAMMALS

EONS	PERIODS
	Quaternary (3 to present)
	Late Tertiary (25 to 3)
	Early Tertiary (65 to 25)
	Cretaceous (135 to 65)
	Jurassic (210 to 135)
Phanerozoic (570 to present)	Triassic (250 to 210)
	Permian (280 to 250)
	Carboniferous (345 to 280)
	Devonian (400 to 345)
	Silurian (435 to 400)
	Ordovician (500 to 435)
	Cambrian (570 to 500)
Proterozoic (2500 to 570)	
Archean (4600 to 2500)	

The Tertiary period, running from 65 to 3 million years ago, is known as the "age of mammals," and because of their great diversity, many more plant and animal species are alive today than at any other time in geologic history. The appearance of the grasses early in the period spawned the evolution of hoofed animals as well as voracious carnivores to prey on them. The prosimians (pre-apes) were also on the scene and gave rise to the anthropoids, ancestors of apes and humans.

Extremes in climate and topography created a greater variety of living conditions than existed during any other equivalent span of geologic time. The rigorous environments presented many challenging opportunities for plants and animals, and the extent to which they invaded diverse habitats was truly remarkable. The Tertiary was a time of constant change, and all species had to adapt to a wide range of living conditions. The changing climate patterns resulted from the movement of continents toward their present positions and from the intense mountain building that raised most ranges of the world.

THE AGE OF MAMMALS

Mammals originated in the late Triassic at roughly the same time as the dinosaurs, and the two groups coexisted for about 150 million years thereafter. When the dinosaurs left the stage at the end of the Cretaceous, the mammals were waiting in the wings, poised to conquer the Earth. Because the dinosaurs represented the largest group of animals, their departure left the world wide open to invasion by the mammals.

About 10 million years after the extinction of the dinosaurs, mammals began to radiate into dazzling arrays of new species. The small, nocturnal mammals eventually evolved into larger animals, some of which were evolutionary dead ends. Of the 30 or so orders of mammals that existed during the early Cenozoic, only half had lived in the preceding Cretaceous while almost two-thirds are still living today.

The evolution of the mammals following the dinosaur extinction was not gradual but progressed in fits and starts. The early Tertiary was characterized by an evolutionary lag, as though the world had not yet awakened

Figure 13-1   Paleogeography (ancient landforms) of the upper Tertiary in North America, when inland seas withdrew from the continents.

from the great extinction. By the end of the Paleocene epoch, about 54 million years ago, mammals began to diversify rapidly. About 37 million years ago, a sharp extinction event took out many of the archaic mammal species, most of which were large, peculiar looking animals. Afterward, the truly modern mammals began to evolve.

The extinction coincided with changes in the deep-ocean circulation and eliminated many species of marine life on the European continent, which was flooded with shallow seas. The separation of Greenland from Europe might have allowed frigid Arctic waters to drain into the North Atlantic, significantly lowering its temperature and causing most types of foraminifera (marine protozoans) to disappear. The climate grew much colder, and the seas withdrew from the land (Fig. 13-1) as the ocean dropped 1,000 feet to perhaps its lowest level of the last several hundred million years.

Much of the drop in sea level might have resulted from the accumulation of massive ice sheets atop Antarctica, which had drifted over the South Pole. A large fall in sea level due to a major expansion of the Antarctic ice sheet led to another extinction about 11 million years ago. These cooling events removed the most vulnerable of species, so that those living today are more robust, having withstood extreme environmental swings over the last 3 million years, when glaciers spanned much of the Northern Hemisphere.

Marine species that survived the great Cretaceous extinction were similar to those that lived in the Mesozoic. Although the extinction in the oceans was severe and many species died out, few radical species appeared because habitats left vacant were simply taken over by the next of kin.

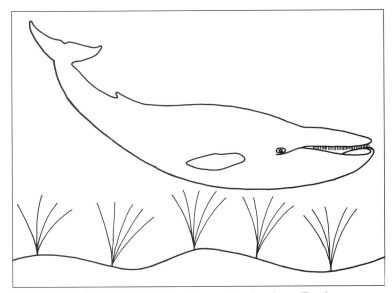

Figure 13-2   Blue whales are the largest animals on Earth.

Species inhabiting unstable environments such as those in the higher latitudes were especially successful.

Some 70 species of marine mammals known as cetaceans were among the most adaptable animals and included dolphins, porpoises, and whales, which evolved during the middle Cenozoic. The ancestors of the whales walked on land and swam in rivers and lakes about 50 million years ago. Today, their closest relatives are the artiodactyls, or hoofed animals with an even number of toes, such as cows, pigs, deer, camels, and giraffes. Ancestors of the blue whale (Fig. 13-2), the largest animal on Earth, evolved from ancient toothed whales about 40 million years ago.

The drifting of continents isolated many groups of mammals, and these evolved along independent lines. For the last 40 million years or so, Australia has been an island continent, without a land link to the other continents. The large island is home to many strange egg-laying mammals called monotremes that include the spiny anteater and the platypus, which should rightfully be classified as surviving mammal-like reptiles. Marsupials, which are primitive mammals that suckle their tiny infants in belly pouches, originated in North America around 100 million years ago, migrated to South America, crossed over to Antarctica when the two continents were still in contact, and landed in Australia before it broke away from Antarctica. Today, 13 of the world's 16 marsupial families reside only in Australia.

The Australian marsupials consist of kangaroos, wombats, and bandicoots, with opossums and related animals occupying other parts of the world. The largest marsupial fossil found is that of the diprotodon, which was about the size of a rhinoceros. The giant kangaroos disappeared soon after early humans invaded the continent some 60,000 years ago. Madagascar, which broke away from Africa about 125 million years ago, has none of the large mammals on the African continent except the hippopotamus, which mysteriously landed on the island after it had drifted some distance from the African mainland.

Camels, which originated about 25 million years ago, migrated out of North America to other parts of the world by connecting land bridges. Horses originated in western North America during the Eocene when they were only about the size of small dogs. As they became progressively larger, their faces and teeth grew longer as the animals switched from browsing to grazing, and their toes fused into hoofs. The giraffes shifted from grazing on grass to browsing on leaves, and their necks lengthened to reach the tall branches. Many types of hoofed animals called ungulates evolved in response to increased grassland all over the world.

All major groups of modern plants were represented in the early Tertiary (Fig. 13-3). The angiosperms dominated the plant world, and all modern families appear to have evolved by about 25 million years ago. Grasses were the most important angiosperms, providing food for ungulates throughout

Figure 13-3   Tertiary plants of the Chickaloon Formation, Cook Inlet region, Alaska. Photo by J. A. Wolfe, courtesy of USGS

the Cenozoic. The grazing habits of many large mammals probably evolved in response to the widespread availability of grasslands.

The first primates lived some 60 million years ago and were about the size of a mouse. Afterward, the primate family tree split into two branches, with monkeys on one limb and the great apes, including our humanlike ancestors, the hominoids, on the other. Beginning about 37 million years ago, the New World monkeys unexplainably migrated to South America from Africa when those continents had already drifted far apart. About 30 million years ago, the precursors of apes lived in the dense tropical rain forests of Egypt, which is now mostly desert. These apelike ancestors migrated from Africa into Europe and Asia between about 25 and 10 million years ago (Fig. 13-4).

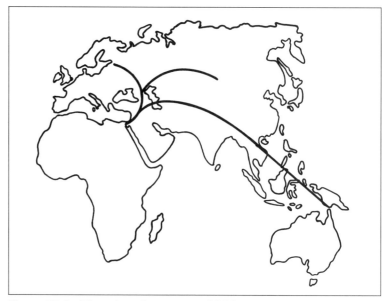

Figure 13-4   The migration routes of African hominoids.

Between 12 and 9 million years ago, the forests of Europe were home for a tree-living, fruit-eating ape called Dryopithecus, which probably evolved into Ramapithecus, an early Asian hominoid with more advanced characteristics than earlier species. Between 9 and 4 million years ago, the fossil record jumps from the hominidlike but mainly ape form of Ramapithecus to the true hominids and the human line of evolution. During this time, much of Africa entered a period of cooler, drier climates and retreating forests, offering many evolutionary challenges to the ancestors of humans.

TERTIARY VOLCANICS

Volcanic activity (Table 13-1) was extensive during the Tertiary. Flood basalts were caused by hot spot volcanism, with plumes of magma rising to the surface from deep within the mantle. India's Deccan Traps (Fig. 13-5) are perhaps the greatest outpouring of basalt on land during the last 250 million years. About 65 million years ago, a giant rift ran down the west side of India, and huge volumes of molten magma poured onto the surface. Some 100 separate flows spilled over 350,000 cubic miles of lava onto much of west-central India, totaling up to 8,000 feet thick over a period of several million years. If spread evenly, that vast amount of lava would envelop the entire world in a layer of volcanic rock some 10 feet thick.

TABLE 13–1   COMPARISON BETWEEN TYPES OF VOLCANISM

Characteristic	Subduction	Rift Zone	Hot Spot
Location	Deep ocean trenches	Midocean ridges	Interior of plates
Percent active volcanoes	80 percent	15 percent	5 percent
Topography	Mountains island arcs	Submarine ridges	Mountains geysers
Examples	Andes Mts. Japan Is.	Azores Is. Iceland	Hawaiian Is. Yellowstone
Heat source	Plate friction	Convection currents	Upwelling from core
Magma temperature	Low	High	Low
Magma viscosity	High	Low	Low
Volatile content	High	Low	Low
Silica content	High	Low	Low
Type of eruption	Explosive	Effusive	Both
Volcanic products	Pyroclasts	Lava	Both
Rock type	Rhyolite Andesite	Basalt	Basalt
Type of cone	Composite	Cinder fissure	Cinder shield

During the eruptions, India was about 300 miles northeast of Madagascar, as it continued drifting toward southern Asia. The Seychelles Bank is a large oceanic volcanic plateau that became separated from the Indian subcontinent and is now exposed on the surface as several islands. The massive outpourings of carbon dioxide-laden lava might have created the extraordinary warm climate of the Paleocene that sparked the evolution of the mammals.

Continental rifting that occurred at approximately the same time as the Deccan Traps eruptions began separating Greenland from Norway and North America. The rifting poured out great flood basalts across eastern Greenland, northwestern Britain, northern Ireland, and the Faeroe Islands, between Britain and Iceland. The island of ice is itself an expression of the Mid-Atlantic Ridge, where massive floods of basalt formed a huge volcanic plateau that rose above sea level about 16 million years ago.

Evidence of substantial explosive volcanism lies in an extensive region from the South Atlantic to Antarctica. The Kerguelen plateau located north

of Antarctica is the world's largest submerged volcanic plateau. It originated from the ocean floor more than 90 million years ago, when a series
of volcanic eruptions released immense quantities of basalt onto the Antarctic plate. The timing also coincides with a mass extinction of species.
The Ninety East Ridge, named for its longitude, 90 degrees east, is an
undersea volcanic mountain range that runs 3,000 miles south from the Bay
of Bengal, India, and formed when the Indian plate passed over a hot spot
as it continued drifting toward Asia.

Prior to the opening of the Red Sea and Gulf of Aden, massive floods of
basalt covered some 300,000 square miles of Ethiopia, beginning about 35
million years ago. The East African Rift Valley extends from the shores of
Mozambique to the Red Sea, where it splits to form the Afar Triangle in
Ethiopia. For the past 25 to 30 million years, the Afar Triangle has been

Figure 13-5   The Deccan Traps flood basalts in India.

Figure 13-6   Columbia River basalt, looking downstream from Palouse Falls, Franklin-Whitman counties in Washington. Photo by F. O. Jones, courtesy of USGS

stewing with volcanism. An expanding mass of molten magma lying just beneath the crust uplifted much of the area thousands of feet.

In North America, major episodes of basalt volcanism occurred in the Columbia River Plateau, the Colorado Plateau, and the Sierra Madre region in northern Mexico. A band of volcanoes stretching from Colorado to Nevada produced a series of very violent eruptions between 30 million and 26 million years ago. Beginning about 17 million years ago and extending over a period of 2 million years, great outpourings of basalt covered Washington, Oregon, and Idaho, creating the Columbia River Plateau (Fig. 13-6). Massive floods of lava enveloped an area of about 200,000 square miles, in places reaching 10,000 feet thick. Periodically, volcanic eruptions spewed out batches of basalt as large as 1,200 cubic miles, forming lava lakes up to 450 miles wide in a matter of days.

The volcanic episodes might be related to the present Yellowstone hot spot, which was beneath the Columbia River Plateau region. The hot spot moved eastward relative to the North American plate and can be traced by following volcanic rocks for 400 miles across Idaho's Snake River Plain. During the last 2 million years, it was responsible for three major episodes of volcanic activity in the vicinity of Yellowstone National Park, Wyoming, which number among the greatest catastrophes of nature.

CENOZOIC MOUNTAIN BUILDING

The Cenozoic is known for its intense mountain building, and the spurt in mountain growth over the past 5 million years might have triggered the Pleistocene ice ages. The Rocky Mountains, extending from Mexico to Canada, heaved upward during the Laramide orogeny from about 80 million to 40 million years ago. During the Miocene epoch, beginning some 25 million years ago, a large portion of western North America uplifted, and the entire Rocky Mountain region rose about a mile above sea level. Great blocks of granite soared high above the surrounding terrain, while to

Figure 13-7 The Grand Canyon of northern Arizona. Photo by D. Carroll, courtesy of USGS

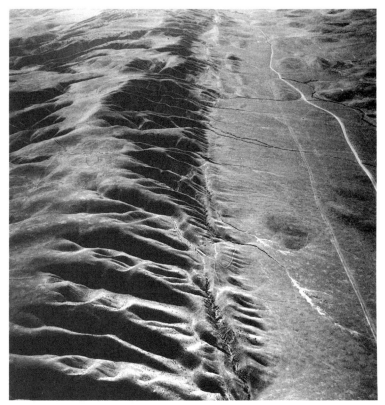

Figure 13-8   The San Andreas Fault in Choice Valley, San Luis Obispo County, California.   Photo R. E. Wallace, courtesy of USGS

the west in the Basin and Range Province the crust stretched and in some places dropped below sea level.

Arizona's Grand Canyon (Fig. 13-7) is at the southwest end of the Colorado plateau, a generally mountain-free expanse that stretches from Arizona north into Utah and east into Colorado and New Mexico. Initially, the area surrounding the canyon was almost totally flat. Over the last 2 billion years, heat and pressure buckled the land into mountains that were later flattened by erosion. Again, mountains formed and eroded, and the region flooded with shallow seas. The land was uplifted another time with the rising of the Rocky Mountains. Between 10 and 20 million years ago, the Colorado River began eroding layers of sediment, exposing the raw basement rock below, and its present course is less than 6 million years old.

About 30 million years ago, the North American continent approached the East Pacific Rise, the counterpart of the Mid-Atlantic Ridge. The first portion of the continent to override the axis of seafloor spreading was the coast of southern California and northwest Mexico. When the rift system and subduction zone converged, the intervening oceanic plate dove into a deep trench. The sediments in the trench were compressed and thrust upward to form California's Coast Ranges. A system of faults associated with the 650-mile-long San Andreas Fault (Fig. 13-8) crisscrosses the mountain belt. The Sierra Nevada Range to the east, which rose about 7,000 feet over the last 10 million years, might be buoyed by a mass of hot rock in the upper mantle.

In the Pacific Northwest of the United States and British Columbia, the Juan de Fuca plate dove into a subduction zone located beneath the continent. As the 50-mile-thick crustal plate subducted into the mantle, its

Figure 13-9   Mount Rainier, Cascade Range, Pierce County, Washington. Photo by B. Willis, courtesy of USGS

heat melted parts of the descending plate and the adjacent lithospheric plate, forming pockets of magma. The magma rose toward the surface, forming the volcanoes of the Cascade Range (Fig. 13-9), which erupted one after another.

India and the rocks that comprise the Himalayas broke away from Gondwana early in the Cretaceous, sped across the ancestral Indian Ocean, and slammed into southern Asia about 45 million years ago. As the Indian and Asian plates collided, the oceanic lithosphere between them thrust under Tibet, destroying 6,000 miles of subducting plate. The increased buoyancy uplifted the Himalaya Mountains and the broad Tibetan Plateau, the equal of which has probably not existed on this planet for over a billion years.

During the past 5 to 10 million years, the entire region rose over a mile in elevation. The continental collision heated vast amounts of carbonate rock, spewing several hundred trillion tons of carbon dioxide into the atmosphere, which might explain why the Earth grew so warm during the Eocene epoch from 54 million to 37 million years ago, when temperatures reached the highest of the past 65 million years. According to the fossil record, winters were warm enough for crocodiles to roam as far north as Wyoming, and forests of palms, cycads, and ferns covered Montana.

About 50 million years ago, the Tethys Sea separating Eurasia from Africa narrowed as the two continents approached each other, then began to close off entirely some 20 million years ago. Thick sediments that had been accumulating for tens of millions of years buckled into long belts of mountain ranges on the northern and southern flanks (Fig. 13-10). The contact between the continents initiated a major mountain building episode that raised the Alps and other ranges in Europe and squeezed out the Tethys Sea.

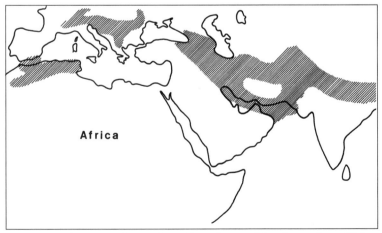

Figure 13-10   Active fold belts in Eurasia, resulting from the collision of lithospheric plates.

This episode of mountain building, called the Alpine orogeny, raised the Pyrenees on the border between Spain and France, the Atlas Mountains of northwest Africa, and the Carpathians in east-central Europe. The Alps of northern Italy formed in much the same manner as the Himalayas, when the Italian prong of the African plate thrust into the European plate.

In South America, the mountainous spine that comprises the Andes and runs along the western edge of the continent rose throughout much of the Cenozoic due to an increase in crustal buoyancy from the subduction of the Nazca plate beneath the South American plate. By the time all the continents had wandered to their present positions and all the mountain ranges had risen to their current heights, the world was ripe for the coming of the ice age.

TERTIARY TECTONICS

Changing climate patterns resulted from the movement of continents toward their present positions, and intense tectonic activity built landforms and raised most mountain ranges of the world. Except for a few land bridges exposed from time to time, plants and animals were prevented from migrating from one continent to the other. About 57 million years ago Greenland began to separate from North America and Eurasia. Prior to about 4 million years ago, Greenland was largely ice-free, but today the world's largest island is buried under a sheet of ice up to 2 miles thick. Alaska connected with east Siberia and closed off the Arctic Basin from warm water currents originating in the tropics, resulting in the formation of pack ice in the Arctic Ocean.

A narrow, curved land bridge that temporarily connected South America with Antarctica assisted in the migration of marsupials to Australia. Sediments containing fossils of a large crocodile, a 6-foot flightless bird, and a 30-foot whale suggest that land bridges existed as late as 40 million years ago. Antarctica and Australia then broke away from South America and moved eastward. When the two continents rifted apart, Antarctica moved toward the South Pole, while Australia continued in a northeastward direction. After Antarctica separated from Australia in the Eocene about 40 million years ago, it drifted over the South Pole, and acquired a permanent ice sheet that buried most of its terrain features (Fig. 13-11).

The Mid-Atlantic Ridge system, which generates new ocean crust in the Atlantic Basin, began to occupy its present location midway between the Americas and Eurasia/Africa about 16 million years ago. Iceland is a broad volcanic plateau of the Mid-Atlantic Ridge that rose above sea level at about the same time. About 3 million years ago, the Panama Isthmus separating North and South America uplifted as oceanic plates collided. Prior to the continental collision, South America had been an island continent for the

Figure 13-11   The Antarctic Peninsula ice plateau, showing mountains literally buried in ice. Photo by P. D. Rowley, courtesy of USGS

past 80 million years, during which time its mammals evolved undisturbed by outside competitors.

A barrier created by the land bridge isolated Atlantic and Pacific species, and extinctions impoverished the once rich fauna of the western Atlantic. The new landform halted the flow of cold water currents from the Atlantic into the Pacific, which along with the closing of the Arctic Ocean from warm Pacific currents might have initiated the Pleistocene glacial epoch. Never before have permanent ice caps existed at both poles, suggesting that the planet has been steadily cooling since the Cretaceous, when average global temperatures were 10 to 15 degrees Celsius warmer than today.

CLOSING OF THE TETHYS

The Tethys Sea was a large, shallow equatorial body of water that separated the southern and northern continents during the Mesozoic and early Cenozoic. About 17 million years ago, the Tethys, which linked the Indian and Atlantic oceans, began to close off as Africa rammed into Eurasia

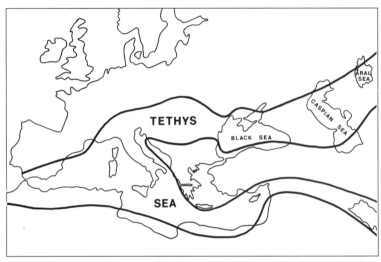

Figure 13-12   The closing of the Tethys Sea by the collision of Africa with Europe and Asia about 20 million years ago.

(Fig. 13-12). The collision also initiated a major mountain building episode that raised the Alps and other ranges. The climates of Europe and Asia were warmer and forests were widespread and lusher than today.

The Mediterranean Basin was apparently cut off from the Atlantic Ocean 6 million years ago, when an isthmus, created at Gibraltar by the northward movement of the African plate, formed a dam across the strait. Nearly a million cubic miles of seawater evaporated, almost completely emptying the basin over a period of about 1,000 years. The adjacent Black Sea, which is 750 miles long and 7,000 feet deep, might have had a similar fate. Like the Mediterranean, it is a remnant of an ancient equatorial sea that separated Africa from Europe.

The collision of the African plate with the Eurasia plate squeezed out the Tethys, resulting in a long chain of mountains and two major inland seas, the ancestral Mediterranean and a composite of the Black, Caspian, and Aral Seas, called the Paratethys, which covered much of Eastern Europe. About 15 million years ago, the Mediterranean separated from the Paratethys, which became a brackish (slightly salty) sea, much like the Black Sea of today.

The disintegration of the great inland waterway was closely associated with the sudden drying of the Mediterranean. The waters of the Black Sea drained into the desiccated basin of the Mediterranean. In a brief moment in geologic time, the Black Sea became practically a dry basin. Then during the last ice age, it refilled again and became a freshwater lake. The brackish and largely stagnant sea that occupies the basin today has evolved since the end of the last ice age.

14

QUATERNARY GLACIATION

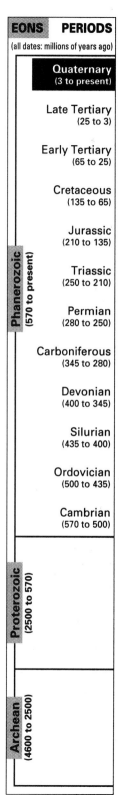

The Quaternary period, from 3 million years ago to the present, witnessed a progression of ice ages that occurred almost like clockwork. The movement of continents to their present locations and the raising of land to higher elevations made geographic conditions ripe for a colder climate. Variations in the Earth's orbital motions might have provided the initial kick to trigger the growth of continental glaciers, which partly explains the recurrence of the ice age cycles about every 100,000 years. Once in place, the glaciers became self-sustaining by controlling the climate to their benefit. Then, mysteriously, in only a few thousand years, the great ice sheets collapsed and rapidly retreated to the poles.

Many northern lands owe their unusual topographies to massive ice sheets that swept down from the polar regions during the Pleistocene epoch of the last 3 million years. The invasion was so pervasive that ice sheets 2 miles or more thick enveloped upper North America and Eurasia, Antarctica, and parts of the Southern Hemisphere. In many areas, the glaciers stripped off entire layers of sediment down to the bare bedrock, erasing the entire geologic history of the region. The last ice age even destroyed evidence of earlier glaciations, as great ice sheets eradicated much of the northern landscape.

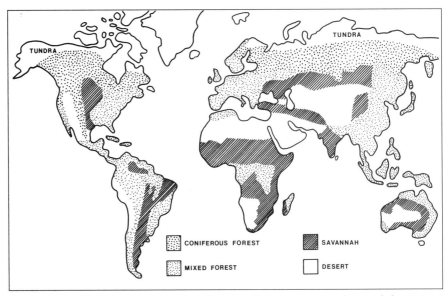

Figure 14-1   World environments, showing forests, savannahs, and deserts as they exist today.

THE AGE OF MAN

About 3 million years ago, huge volcanic eruptions in the northern Pacific Ocean darkened the skies and global temperatures plummeted, culminating in a series of glacial episodes. The climate change prompted a shift from forested environments to open savanna habitats in Africa (Fig. 14-1). These changing conditions produced many new animal species and spurred the evolution of early humans. Indeed, we are products of the ice ages, which spanned the whole of human experience.

Our direct ancestors the hominids evolved in Africa, probably from the same species that gave rise to the great apes, including the gorilla and chimpanzee. Around 7 million years ago, much of Africa entered a period of cooler, drier climates when forests retreated and were replaced with grasslands. Life on the savanna, where our ancient ancestors roamed, was harsher and more challenging than life in the forests, where the apes lived. To survive under these difficult conditions, early humans rapidly evolved into intelligent, upright-walking species, whereas the apes are much the same today as they were millions of years ago.

An early hominid species called *Australopithecus* first appeared in Africa about 4 million years ago. It walked on two legs but retained many apelike features, such as long arms in relation to its legs and curved bones in its hands and feet. The species was quite muscular and considerably stronger than modern humans. Males stood a little less than 5 feet tall and weighed about 100 pounds, and females stood about 4 feet tall and weighed about 70 pounds. Two or more lines of Australopithecene lived simultaneously in Africa and survived practically unchanged for more than a million years. After a lengthy period of apparent stability, all but one line became extinct, possibly due to a changing climate or habitat.

The most successful of the early hominids was *Homo habilis,* which evolved a little over 2 million years ago. It was a transitional species, somewhere between primitive apelike hominids and humans. Its brain was about half the size of a modern human brain. The limb bones were markedly different from earlier hominids and more closely resembled those of later humans. It was the first human species to make and use tools and had a well-developed speech center, indicating a primitive language capability.

Homo habilis disappeared from Africa around 1.8 million years ago and was replaced by *Homo erectus.* This species is widely accepted as human and appears to have evolved in Africa directly from *Homo habilis.* It also

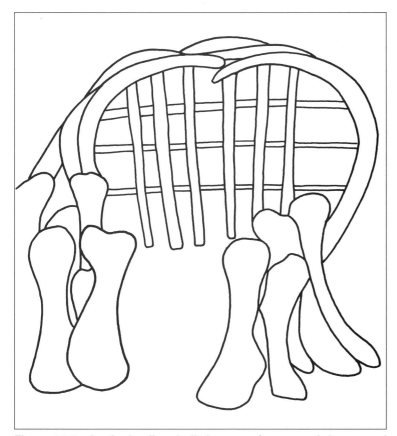

Figure 14-2 Arctic dwellers built houses of mammoth bones and tusks covered with hides.

could have evolved independently in Asia and subsequently migrated to Africa. About a million years ago, this early human occupied southern and eastern Asia, where it lived until about 200,000 years ago. Its advanced features suggest a spurt of evolutionary development, as it shares many attributes with modern humans. Its brain was also larger, about two-thirds the size of a modern human's.

Many types of *Homo erectus* then scattered throughout the world, which suggests that anatomically modern humans evolved from this species in several places, possibly accounting for the differences in races among people today. Peking man, a variety of *Homo erectus* that lived in China about 400,000 years ago, dwelled in caves and was possibly the first to use fire. Another variety, called Java man, arrived in Java about 700,000 years ago. About 60,000 years ago, the descendants of Java man migrated to Australia and possibly to the South Pacific islands.

The earliest *Homo sapiens,* called Cro-Magnon for the Cro-Magnon cave in France where the first discoveries were made in 1868, originated in Africa perhaps 200,000 years ago. Evidence also suggests they arose simultaneously in several parts of the world, as much as a million years ago, possibly evolving directly from *Homo erectus.* The Cro-Magnon shared most of the physical attributes of modern humans. The skull, whose brain case proportions were modern in structure, was short, high, and rounded, and the lower jaw ended in a definite chin. The rest of the skeleton was slender and long-limbed compared to earlier species of *Homo.*

Sometime during the last ice age, Cro-Magnon appears to have advanced into Europe and Asia during a warm interlude when the climate was less severe. They probably lived much like present-day natives of the Arctic tundra, fishing the rivers and hunting reindeer and other animals. Due to the scarcity of wood in the cold tundra, ice age hunters on the central Russian plain built houses of mammoth bones and tusks covered with animal hides (Fig. 14-2) and burned bones and animal fat for heat and light.

The Neanderthals were primitive *Homo sapiens,* named for the Neander Valley near Düsseldorf, Germany, where the first fossils were recognized in 1856. They are generally thought to have inhabited caves, but they also occupied open-air sites, as evidenced by hearths and rings made of mammoth bones and stone tools normally associated with these people. During the last interglacial period, called the Eemian, which ran from about 135,000 to 115,000 years ago, the Neanderthals ranged over most of western Europe and central Asia, extending as far north as the Arctic Ocean.

Modern humans and Neanderthals apparently coexisted in Eurasia for at least 60,000 years and shared many cultural advancements. The Neanderthals thrived in these regions until about 30,000 years ago, declining over a period of perhaps 5,000 years. The disappearance of the Neanderthals might have resulted from their replacement or assimilation by modern humans.

THE PLEISTOCENE ICE AGES

From the end of the Permian 250 million years ago to about 40 million years ago, no major ice caps covered the world, suggesting that the existence of ice sheets at both poles during the Pleistocene was a unique event in Earth history. Analysis of deep-sea sediments and glacial ice cores provides a historical record of the recent ice ages, which began about 3 million years ago, when a progression of glaciers sprawled across the northern continents.

During this time, the surface waters of the ocean cooled dramatically, causing diatoms, a species of algae with shells made of silica (Fig. 14-3), to sharply decline in the Antarctic, presumably when sea ice reached its maximum northern extent, thus shading the algae below. In the absence of sunlight for photosynthesis, the diatoms vanished, and their disappearance marks the initiation of the Pleistocene glacial epoch in the Northern Hemisphere.

The latest period of glaciation began over 100,000 years ago, intensified about 75,000 years ago, possibly due to the massive Mount Toba eruption

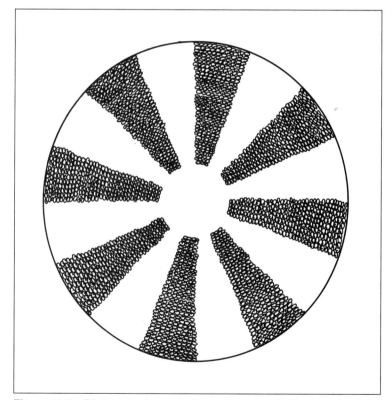

Figure 14-3   Diatoms, whose shells were made of silica, survived the end-Cretaceous extinctions.

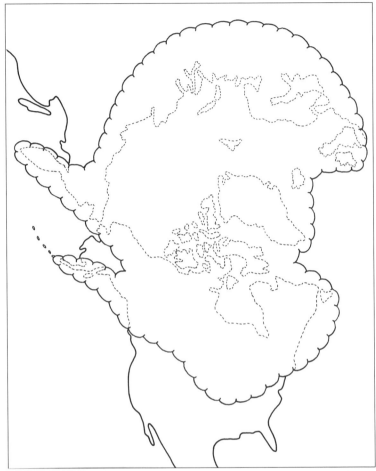

Figure 14-4 Extent of the glaciers during the last ice age, as shown by billowy outline.

in Indonesia, and peaked about 18,000 years ago. During the height of the last ice age, glaciers up to 2 or more miles thick enveloped Canada, Greenland, and Northern Europe (Fig. 14-4).

In North America, the largest ice sheet, called the Laurentide, blanketed an area of 5 million square miles. It extended from Hudson Bay, reaching northward into the Arctic Ocean and southward into eastern Canada, New England, and the upper midwestern United States. A smaller ice sheet called the Cordilleran originated in the Canadian Rockies and engulfed western Canada, along with the northern and southern parts of Alaska, leaving an ice-free corridor down the middle, used by humans migrating into North America. Glaciers also covered small portions of the northwestern United States.

The largest ice sheet in Europe, the Fennoscandian, fanned out from northern Scandinavia and covered most of Great Britain as far south as London and large parts of northern Germany, Poland, and European Russia. A smaller ice sheet called the Alpine, centered in the Swiss Alps, covered parts of Austria, Italy, France, and southern Germany. In Asia, ice sheets draped over the Himalayas and blanketed parts of Siberia.

In the Southern Hemisphere, small ice sheets capped the mountains of Australia, New Zealand, and the Andes of South America. Elsewhere, alpine glaciers topped mountains that are presently ice free. Only Antarctica had a major ice sheet, which expanded to about 10 percent larger than its present size, extending as far as the southern tip of South America.

Excess ice with nowhere to go except into the sea calved off to form icebergs. During the peak of the last ice age, icebergs covered half the area of the ocean. The ice floating in the sea reflected sunlight back into space, thereby maintaining a cool climate with average global temperatures about 5 degrees Celsius colder than today.

Some 10 million cubic miles of water were tied up in the continental ice sheets, which covered about a third of the land surface with an ice volume three times greater than its present size. The accumulated ice dropped the level of the ocean about 400 feet, advancing the shoreline tens of miles

TABLE 14–1   MAJOR DESERTS OF THE WORLD

Desert	Location	Type	Area (square miles × 1000)
Sahara	North Africa	Tropical	3500
Australian	Western/Interior	Tropical	1300
Arabian	Arabian Peninsula	Tropical	1000
Turkmenistan	S. Central Asia	Continental	750
North America	S.W. U.S./ N. Mexico	Continental	500
Patagonian	Argentina	Continental	260
Thar	India/Pakistan	Tropical	230
Kalahari	S.W. Africa	Littoral	220
Gobi	Mongolia/China	Continental	200
Takla Makan	Sinkiang, China	Continental	200
Iranian	Iran/Afganistan	Tropical	150
Atacama	Peru/Chile	Littoral	140

Figure 14-5   Antonelli Glacier, showing rugged periglacial area, including recessional and other moraines. Photo by M. F. Meier, courtesy of USGS

seaward. The drop in sea level exposed land bridges and linked continents, spurring the migration of species to various parts of the world. The great weight of the ice sheets caused the continental crust to sink deeper into the upper mantle. Even today, the northern lands are rebounding as much as half an inch per year, long after the weight of the glaciers was lifted.

The lower temperatures reduced the evaporation rate of seawater and lowered the average amount of precipitation, which caused the expansion of deserts in many parts of the world (Table 14-1). Fierce desert winds produced tremendous dust storms, and the dense dust suspended in the atmosphere blocked sunlight, dropping temperatures even further. Most of the windblown sand deposits called loess in the central United States were laid down during the Pleistocene ice ages.

The cold weather and approaching ice forced species to migrate to warmer latitudes. Ahead of the ice sheets, which advanced on average perhaps a few hundred feet per year, lush deciduous woodlands gave way to evergreen forests that yielded to grasslands, which became barren tundra and rugged periglacial regions on the margins of the ice sheets (Fig. 14-5).

THE HOLOCENE INTERGLACIAL

Perhaps one of the most dramatic climate changes in geologic history took place during the present interglacial period, known as the Holocene epoch. After some 100,000 years of gradual accumulation of snow and ice up to 2 miles and more thick, the glaciers melted away in only a few thousand years, retreating upwards of a half-mile annually. About a third of the ice melted between 16,000 and 12,000 years ago, when average global temperatures increased about 5 degrees Celsius to nearly present-day levels. A renewal of the deep-ocean circulation system, which was shut off or weakened during the ice age, might have thawed out the planet from the deep freeze.

A gigantic ice dam on the border of Idaho and Montana held back a huge lake hundreds of miles wide and up to 2,000 feet deep. Around 13,000 years ago, the sudden bursting of the dam sent waters gushing toward the Pacific Ocean. Along the way, the floodwaters carved out one of the strangest landscapes known, called the Scablands (Fig. 14-6). Lake Agassiz, a vast reservoir of meltwater formed in a bedrock depression carved out by glaciers, sat at the edge of the retreating ice sheet in southern Manitoba, Canada.

Figure 14-6　Palouse Island near the eastern margin of the Scablands, Washington. Photo by F. O. Jones, courtesy of USGS

When the North American ice sheet retreated, its meltwaters flowed down the Mississippi River into the Gulf of Mexico. After the ice sheet subsided beyond the Great Lakes, the meltwater took an alternate route down the St. Lawrence River, and the cold waters entered the North Atlantic Ocean. Simultaneously, the Niagara River Falls began cutting its gorge and has traversed more than 5 miles northward since the ice sheet melted.

The rapid melting of the glaciers culminated in the extinction of microscopic organisms called foraminifera (Fig. 14-7), which met their demise when a torrent of meltwater and icebergs spilled into the North Atlantic. The massive floods formed a cold fresh water lid on the ocean that significantly changed the salinity of the seawater. The cold waters also blocked poleward-flowing warm currents from the tropics, causing land temperature to fall to near ice age levels.

The ice sheets appear to have paused in mid-stride between 13,000 and 11,500 years ago, during a period called the Younger Dryas, named for an Arctic wildflower that grew in Europe. Afterward, the warm currents returned, and the mild weather remained, prompting a second episode of melting that led to the present volume of ice by about 6,000 years ago.

Figure 14-7   Foraminifera were important limestone builders.

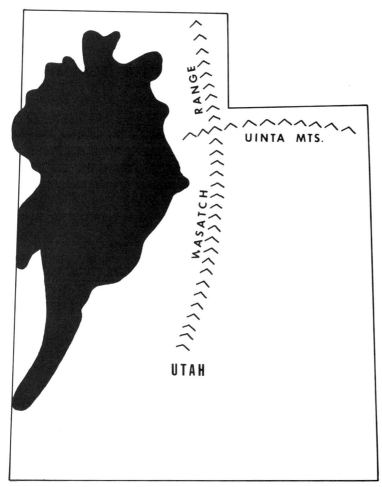

Figure 14-8 The approximate extent of the ancestral Great Salt Lake, from about 12,000 to 6,000 years ago.

Following the receding ice sheets, plants and animals began to return to the northern latitudes.

When the ice sheets melted, massive floods raged across the land as water gushed from trapped reservoirs below the glaciers. While flowing under the ice, water surged in vast turbulent sheets that scoured deep grooves in the crust, forming steep ridges carved out of solid bedrock. Each flood continued until the weight of the ice sheet shut off the outlet of the reservoir. When water pressure built up again, another massive surge of meltwater spouted from beneath the glacier and rushed toward the sea. Huge torrents of meltwater laden with sediment surged along the Mississippi River toward the Gulf of Mexico, widening the channel to several times its present size. Many other rivers overreached their banks to carve out new floodplains.

The warming paved the way for the Climatic Optimum, a period of unusually warm, wet conditions that began 6,000 years ago and lasted for 2,000 years. As the Climatic Optimum unfolded, many regions of the world warmed on average about 5 degrees Celsius. The melting ice caps released massive floodwaters into the sea, raising sea levels 300 feet higher than when the Holocene began.

The inland seas filled with sediments, and subsequent uplifting drove out the waters, leaving behind salt lakes. Great Salt Lake in Utah is today only a remnant of a vast inland sea. During a long wet period between 12,000 and 6,000 years ago, it expanded to several times its current size and flooded the nearby salt flats (Fig. 14-8).

MEGAHERBIVORE EXTINCTION

When the last ice age was drawing to a close between 12,000 and 10,000 years ago, an unusual extinction killed off large terrestrial plant-eating mammals called megaherbivores. Woolly rhinos, mammoths, and Irish elk disappeared in Eurasia, and the great buffalo, giant hartebeests, and giant horses disappeared in Africa. Over 80 percent of the large mammals and a significant number of bird species disappeared from Australia.

Meanwhile, the giant ground sloths, mastodons, and woolly mammoths disappeared from North America. A possible exception was the dwarf woolly mammoth, which might have survived in the Arctic until about 4,000 years ago. The loss of these animals also caused their main predators the American lion, saber-toothed tiger, and dire wolf to go extinct.

The global environment reacted to the changing climate at the end of the last ice age with declining forests and expanding grasslands. The climate change disrupted the food chains of many large animals; deprived of their food resources they simply vanished. Also by this time, humans were becoming proficient hunters and roamed northward on the heels of the retreating glaciers. On their journey, they encountered an abundance of wildlife, many species of which they might have hunted to extinction.

In North America, 35 classes of mammals and 10 classes of birds went extinct. The extinctions occurred between 13,000 and 10,000 years ago, with the greatest die-out peaking around 11,000 years ago. Most mammals adversely affected were large herbivores weighing over 100 pounds, many of them weighing up to a ton or more. Unlike earlier episodes of mass extinction, this event did not significantly affect small mammals, amphibians, reptiles, and marine invertebrates. Strangely, after having endured several previous periods of glaciation and deglaciation over the past 2 to 3 million years, these large mammals were unable to survive at the end of the last ice age.

During this time, ice age peoples occupied many parts of North America, and their spearpoints have been found resting among the remains of giant mammals, including mammoths, mastodons, tapirs, native horses, and camels. These people crossed into North America from Asia over a land bridge formed by the draining of the Bering Sea and moved through an ice-free corridor east of the Canadian Rockies.

Instead of migrating to North America in several waves, small bands of nomadic hunters probably crossed the ancient land bridges and ended up in the New World purely by accident. The human hunters arriving from Asia sped across the virgin continent following migrating herds of large herbivores, leaving big game carcasses in their wake.

GLACIAL GEOLOGY

Much of the landscape in the northern latitudes owes its unusual geography to massive glaciers that swept down from the polar regions during the last ice age. The power of glacial erosion is well demonstrated by deep-sided valleys carved out of mountain slopes by thick sheets of flowing ice (Fig. 14-9) a mile or more thick. The glaciers descended from the mountains and spread across most of the northern lands, destroying everything in their path.

In the alpine regions, glaciers flowing down mountain peaks gouged large pits called cirques (Fig. 14-10). The glaciers extended far down the valleys,

Figure 14-9 A glacial valley, Dolores County, Colorado. Photo by W. Cross, courtesy of USGS

Figure 14-10 A cirque carved by a glacier on Jack Mountain, Cascade Range, Washington. Photo by A. Post, courtesy of USGS

grinding rocks on the valley floors as the ice advanced and receded. In effect, a river of solid ice embedded with rocks moved along the valley floors, grinding them down like a giant file as the glacier flowed back and forth over them.

The advancing glaciers left parallel furrows called glacial striae on the valley floors, as they sliced down mountainsides. Miles from existing glaciers are large areas of polished and deeply furrowed rocks, and heaped rocks called moraines mark the extent of former glaciers. Many of the northern lands are dotted with glacial lakes developed from deep pits excavated by roving glaciers.

In other areas, older rocks were buried under thick deposits of glacial till, forming elongated hillocks aligned in the same direction called drumlins. They are tall and narrow at the upstream end of the glacier and slope to a low, broad tail. The hills appear in concentrated fields in North America, Scandinavia, Britain, and other areas once covered with ice. Drumlin fields might contain as many as 10,000 knolls, looking much like row upon row of eggs lying on their sides.

Figure 14-11   A glacial boulder field, Tulare County, California. Photo by F. E. Matthes, courtesy of USGS

Eskers are long, sinuous sand deposits produced by glacial outwash stream debris. They form winding, steep-walled ridges that extend up to 500 miles in length but seldom exceed more than 2,000 feet wide and 150 feet high. Eskers were probably created by streams running through tunnels beneath the ice sheet. When the ice melted, the old stream deposits remained standing as a ridge. Well-known esker areas exist in Maine, Canada, Sweden, and Ireland.

Rugged periglacial regions existed at the margins of the ice sheets. Periglacial processes sculpted features along the tip of the ice that were directly controlled by the glacier. Cold winds whistling off the ice sheets influenced the climate of the glacial margins and helped create periglacial conditions. The zone was dominated by such processes as frost heaving, frost splitting, and sorting, creating immense boulder fields from once solid bedrock (Fig. 14-11).

The ice age is still with us, only we are fortunate to live in a warmer period between glaciations. Perhaps in another couple thousand years, massive ice sheets will again be on the rampage, wiping out everything they come across. Rubble from demolished northern cities would be bulldozed hundreds of miles to the south, gripped in the frozen jaws of the advancing ice. Global temperatures would plummet, as plants and animals, including humans, scramble for the warmth of the tropics.

GLOSSARY

abrasion   erosion by friction, generally caused by rock particles
                    carried by running water, ice, and wind

abyss   the deep ocean, generally over a mile in depth

accretion   the accumulation of celestial gas and dust into a
                    planetesimal, asteroid, moon, or planet by gravitational
                    attraction

aerosol   a mass made of solid or liquid particles dispersed in air

age   a geological time interval that is smaller than an epoch

agglomerate   a pyroclastic rock composed of consolidated volcanic
                    fragments

albedo   the amount of sunlight reflected from an object; depend-
                    ent on color and texture

alluvium   stream-deposited sediment

alpine glacier   a mountain glacier or a glacier in a mountain valley

amphibian   a cold-blooded, four-footed vertebrate midway in evolu-
                    tionary development between fishes and reptiles

andesite   a volcanic rock intermediate in characteristics between
                    basalt and rhyolite

angiosperms   flowering plants that reproduce sexually with seeds

annelids	wormlike invertebrates characterized by segmented bodies with distinct heads and appendages
arthropods	the largest group of invertebrates, including crustaceans and insects; characterized by segmented bodies, jointed appendages, and exoskeletons
ash fall	the fallout of small, solid particles from a volcanic eruption cloud
asteroid	a rocky or metallic body orbiting the sun between Mars and Jupiter and possibly once part of a larger body that disintegrated
asteroid belt	a band of asteroids orbiting the sun between the orbits of Mars and Jupiter
asthenosphere	a layer of the upper mantle, roughly between 50 and 200 miles below the surface, that is more plastic than the rock above and below and might be in convective motion
astrobleme	eroded remains of an ancient impact structure produced by a large cosmic body impacting the Earth's surface
atmospheric pressure	the weight per unit area of the total mass of air above a given point; also called barometric pressure
back-arc basin	a seafloor-spreading system of volcanoes caused by extension behind an island arc above a subduction zone. It is a depression in the oceanic crust caused by the effects of plate subduction.
barrier island	a low, elongated coastal island that parallels the shoreline and protects the beach from storms
basalt	a dark volcanic rock that is usually quite fluid in the molten state
basement rock	subterranean igneous, metamorphic, granitized, or highly deformed rock underlying younger sediments
batholith	the largest of the intrusive igneous bodies and more than 40 square miles on its uppermost surface
bedrock	solid layers of rock lying beneath younger material
bicarbonate	an ion created by the action of carbonic acid on surface rocks. Marine organisms use the bicarbonate along with

calcium to build supporting structures composed of calcium carbonate.

big bang   a theory for the creation of the universe, dealing with the initiation of all matter

biogenic   describes sediments composed of the remains of plant and animal life such as shells

biomass   the total mass of living organisms within a specific habitat

biosphere   the living portion of the Earth that interacts with all other biological and geological processes

bivalves   mollusks with a shell comprising two hinged valves; including oysters, muscles, and clams

black smoker   superheated hydrothermal water rising to the surface at a midocean ridge. The water is supersaturated with metals, and when exiting through the seafloor it quickly cools and the dissolved metals precipitate, resulting in black, smokelike effluent.

brachiopods   marine, shallow-water invertebrates with bivalve shells similar to mollusks; plentiful in the Paleozoic

bryozoans   marine invertebrates, characterized by a branching or fanlike structure, that grow in colonies

calcite   a mineral composed of calcium carbonate

caldera   a large pitlike depression at the summits of some volcanoes and formed by great explosive activity and collapse

calving   formation of icebergs by glaciers breaking off and entering the ocean

Cambrian explosion   a rapid radiation of species that occurred as a result of the availability of a large adaptive space, including a large number of habitats and mild climate

carbonaceous   describes a substance containing carbon, namely sedimentary rocks such as limestone and certain types of meteorites

carbonate   a mineral containing calcium carbonate, such as limestone or dolostone

carbon cycle	the flow of carbon into the atmosphere and ocean, the conversion to carbonate rock, and the return into the atmosphere by volcanoes
catastrophism	a theory proposing that recurrent, violent global events cause the sudden disappearance and appearance of species
Cenozoic	an era of geologic time comprising the last 65 million years
cephalopod	marine mollusks, including squid, octopus, cuttlefish, nautilus, and extinct ammonites
chalk	a soft form of limestone composed chiefly of the calcite shells of microorganisms
circum-Pacific belt	an area of active seismic regions around the rim of the Pacific plate, one of which is the Ring of Fire
cirque	a glacial erosional feature, producing an amphitheater-like head of a glacial valley
coal	a fossil fuel deposit originating from metamorphosed plant material
coelenterates	multicellular marine organisms, including jellyfish and corals
comet	a celestial body believed to originate from a cloud of such bodies that surrounds the sun; develops a long tail of gas and dust particles when traveling near the inner Solar System
conglomerate	a sedimentary rock composed of welded fine-grain and coarse-grain rock fragments
continent	a landmass composed of light, granitic rock that rides on denser rocks of the upper mantle
continental drift	the concept that the continents have been moving across the surface of the Earth throughout geologic time
continental glacier	a glacial ice sheet covering a portion of a continent
continental margin	the area in the ocean between the shoreline and the abyss

continental shelf	the offshore area of a continent in shallow seas
continental shield	ancient crustal rocks upon which the continents grew
continental slope	the transition from the continental margin to the deep-sea basin
convection	the circular, vertical flow of a fluid medium by heating from below. As materials are heated, they become less dense and rise, then cool down, become more dense and sink.
coral	any of a large group of shallow-water, bottom-dwelling marine invertebrates that are reef-building colonies common in warm waters
cordillera	parallel chains of mountain ranges; one such grouping is the Rockies, Cascades, and Sierra Nevada in North America and the Andes in South America
core	the central part of the Earth, consisting of a heavy iron-nickel alloy
correlation	the tracing of equivalent rock exposures over distance, usually with the aid of fossils
craton	the stable interior of a continent, usually composed of the oldest rocks
crevasse	a deep fissure in the crust or in a glacier
crinoid	an echinoderm with a calyx including body organs atop a long stalk anchored to the sea floor; also called "sea lily"
crust	the outer layers of a planet, or a moon's rocks
crustaceans	arthropods characterized by two pairs of antennalike appendages forward of the mouth and three pairs behind it; includes shrimp, crabs, and lobsters
diapir	the buoyant rise of a molten rock through heavier rock
diatoms	microplants whose fossil shells form siliceous sediments called diatomaceous earth
divergent plate	the boundary between lithospheric plates. Generally corresponds to midocean ridges where new crust is

	formed by the solidification of liquid rock rising from below
dolostone	a rock formed when calcite in limestone is replaced with magnesium
drumlin	a hill of glacial debris facing in the direction of glacial movement
dune	a ridge of wind-blown sediments, usually in motion
earthquake	the sudden rupture of rocks along active faults in response to geological forces within the Earth
East Pacific Rise	a midocean ridge spreading system running north-south along the eastern side of the Pacific. The predominant location where hot springs and black smokers were discovered.
echinoderms	marine invertebrates, including starfish, sea urchins, and sea cucumbers
ecliptic	the plane of the Earth's orbit around the sun
ecology	the interrelationships between organisms and their environment
ecosphere	the complex interconnections among the biosphere, hydrosphere, atmosphere, and lithosphere
ecosystem	a community of organisms and their environment functioning as a complete, self-contained biological unit
environment	the complex physical and biological factors that act on an organism to determine its survival and evolution
eolian	describes a deposit of wind-blown sediment
eon	the longest unit of geologic time, roughly a billion years or more in duration
epoch	a geologic time unit shorter than a period and longer than an age
era	a unit of geologic time shorter than an eon and consisting of several periods

erosion	the wearing away of surface materials by natural agents such as wind and water
erratic	a glacially transported boulder that has been deposited far from its source
escarpment	a mountain wall caused by the elevation of a block of crust
esker	a curved ridge of glacially deposited material
eukaryote	a highly developed organism with a nucleus that divides genetic material in a systematic manner
evaporite	the deposition of salt, anhydrite, and gypsum from the evaporation of stranded seawater in an enclosed basin
evolution	the changing of physical and biological factors over a period of time
exoskeleton	the hard outer protective covering of invertebrates
extinction	the loss of large numbers of species over a short geologic time
extrusive	an igneous volcanic rock ejected onto the surface of a planet or moon
fault	a break in crustal rocks caused by Earth movements
fauna	the animal life of an area or age
fissure	a large crack in the crust through which magma might escape from a volcano
fluvial	pertaining to river deposits
foraminifera	calcium carbonate-secreting organisms that live in the surface waters of the oceans. After death their shells form the primary constituents of limestone and sediments deposited on the sea floor.
formation	a combination of rock units that can be traced over distance
fossil	any remains, impression, or trace in rock of a plant or animal of a previous geologic age

fossil fuel	an energy source derived from ancient plant and animal life; includes coal, oil, and natural gas. When ignited, these fuels release carbon dioxide that was stored in the Earth's crust for millions of years.
fractionation	a process by which a subducted slab of crustal rocks starts melting on its way down into the mantle, followed by the lighter components rising back to the surface before the heavier ones
fumarole	a vent, such as a geyser, through which steam or other hot gases escape from underground
galaxy	a large gravitationally bound cluster of stars
gastropod	a large class of mollusks, including slugs and snails, characterized by a body protected by a single shell that is often coiled
geothermal	describes hot water or steam generated by hot rocks in the Earth's interior
geyser	a spring that ejects intermittent jets of steam and hot water
glacier	a thick mass of moving ice occurring where winter snowfall exceeds summer melting
glossopteris	a late Paleozoic plant that existed on the southern continents but has not been found on the northern continents, thereby confirming the existence of Gondwana
Gondwana	a southern supercontinent of Paleozoic time, comprised of Africa, South America, India, Australia, and Antarctica. It broke up into the present continents during the Mesozoic era.
graben	a valley formed by a downdropped fault block
granite	a coarse-grained, silica-rich rock, consisting primarily of quartz and feldspars. It is the principal constituent of the continents and is believed to derive from a molten state beneath the Earth's surface.
greenhouse effect	the trapping of heat in the lower atmosphere, principally by water vapor and carbon dioxide
greenstone belt	a mass of Archean metamorphosed igneous rock

guyot	an undersea volcano that once was above sea level and had its top flattened by erosion. Later, subsidence caused the volcano to sink below the ocean surface, preserving its flat-top appearance.
heat budget	the flow of solar energy through the biosphere
horn	a peak on a mountain formed by glacial erosion
horst	an elongated, uplifted block of crust bounded by faults
hot spot	a volcanic center with no relation to a plate boundary; an anomalous magma generation site in the mantle
hydrocarbon	a molecule consisting of carbon chains with attached hydrogen atoms
hydrologic cycle	the flow of water from the ocean to the land and back to the sea
hydrosphere	the water layer at the Earth's surface
hydrothermal	relating to the movement of hot water through the crust
Iapetus Sea	a former sea that occupied a similar area as the present Atlantic Ocean prior to the assemblage of Pangaea
ice age	a period of time when large areas of the Earth were covered by massive glaciers
iceberg	a portion of a glacier calved off upon entering the sea
ice cap	a polar cover of ice and snow
igneous rocks	all rocks solidified from a molten state
impact	the point on the surface upon which a celestial object lands
index fossil	a representative fossil that identifies the rock strata in which it is found
interglacial	a warming period between glacial periods
intrusive	any igneous body that has solidified in place below the Earth's surface
invertebrates	animals with external skeletons such as shellfish and insects

iridium	a rare isotope of platinum, relatively abundant on meteorites
island arc volcanoes	volcanoes landward of a subduction zone that is parallel to a trench and above the melting zone of a subducting plate
isostasy	a geologic principle that states that the Earth's crust is buoyant and rises and sinks depending on its density
karst	a terrain comprised of numerous sinkholes in limestone
Laurasia	a northern supercontinent of Paleozoic time consisting of North America, Europe, and Asia
lava	molten magma that flows out onto the surface
limestone	a sedimentary rock consisting mostly of calcite from shells of marine invertebrates
lithosphere	the rocky outer layer of the mantle that includes the terrestrial and oceanic crusts. The lithosphere circulates between the Earth's surface and mantle by convection currents.
lithospheric plate	describes a segment of the lithosphere involved in the plate interaction of other plates in tectonic activity
loess	a thick deposit of airborne dust
lystrosaurus	an ancient, extinct, mammal-like reptile
magma	a molten rock material generated within the Earth; it is the constituent of igneous rocks
magnetic field reversal	a reversal of the north-south polarity of the magnetic poles
mantle	the part of a planet below the crust and above the core; composed of dense rocks that might be in convective flow
maria	dark plains on the lunar surface produced by massive basalt floods
megaherbivore	a large plant-eating animal such as an elephant or the extinct mastodon

megaplume	a large volume of mineral-rich warm water above an oceanic rift
Mesozoic	literally, the period of middle life, referring to a period between 250 million and 65 million years ago
metamorphism	recrystallization of previous igneous or sedimentary rocks under conditions of intense temperatures and pressures without melting
metazoans	primitive multicellular animals with cells differentiated for specific functions
meteorite	a metallic or stony celestial body that enters the Earth's atmosphere and impacts on the surface
methane	a hydrocarbon gas liberated by the decomposition of organic matter; a major constituent of natural gas
Mid-Atlantic Ridge	the seafloor spreading ridge that marks the extensional edge of the North and South American plates to the west and the Eurasian and African plates to the east
midocean ridge	a submarine ridge along a divergent plate boundary where a new ocean floor is created by the upwelling of mantle material
mollusks	a large group of invertebrates, including snails, clams, squids, and extinct ammonites, many of which are characterized by a hard external shell surrounding the body
moraine	a ridge of erosional debris deposited by the melting margin of a glacier
nebula	an extended astronomical object with a cloudlike appearance. Some nebulae are galaxies; others are clouds of dust and gas within our galaxy.
nutrient	a food substance that nourishes living organisms
Oort cloud	a collection of comets that surrounds the sun about a light-year away
ophiolites	masses of oceanic crust thrust onto the continents by plate collisions; also contain ore deposits

ore body	the accumulation of metal-bearing ores where hot hydrothermal water moving upward toward the surface mixes with cold seawater penetrating downward
orogens	eroded roots of ancient mountain ranges
orogeny	a process of mountain building by tectonic activity
outgassing	the loss of gas from within a planet as opposed to degassing, the loss of gas from meteorites
overthrust	a thrust fault whereby one segment of crust overrides another segment of crust over long distances
ozone	a molecule consisting of three atoms of oxygen; in the upper atmosphere it filters out harmful ultraviolet radiation from the sun
paleomagnetism	the study of the Earth's magnetic field, including the past position and polarity of the poles
paleontology	the study of ancient life forms, based on the fossil record of plants and animals
Paleozoic	the period of ancient life, between 570 million and 250 million years ago
Pangaea	ancient supercontinent that included all the lands of the Earth
Panthalassa	a great world ocean that surrounded Pangaea
pelecypods	aquatic mollusks, including oysters, clams, scallops, and mussels
peridotite	the most common ultramafic (rich in magnesium and iron, very low in silica) rock type in the mantle
period	a division of geologic time longer than an epoch and shorter than an era
permafrost	permanently frozen ground in the Arctic regions
photosynthesis	the process by which plants create carbohydrates from carbon dioxide, water, and sunlight
phyla	groups of organisms that share similar body forms

phytoplankton	marine or freshwater, microscopic, single-celled, freely drifting plant life
pillow lava	lava extruded on the ocean floor and giving rise to tubular shapes
placer	a deposit of rocks left behind by a melting glacier; any ore deposit that is enriched by stream action
planetesimals	small celestial bodies that accreted during the early stages of the Solar System
plate tectonics	the theory that accounts for the major features of the Earth's surface in terms of the interaction of lithospheric plates
pluton	an underground body of igneous rock younger than the rocks that surround it. It is formed where molten rock melts and assimilates older rocks.
podia	structures resembling feet
prebiotic	describes conditions on the early Earth prior to the introduction of life processes
precipitation	products of condensation that fall from clouds as rain, snow, hail, or drizzle; also the deposition of minerals from seawater
primordial	pertaining to the primitive conditions that existed during early stages of the Earth's development
prokaryote	a primitive organism lacking a nucleus
protistids	unicellular organisms, including bacteria, protozoans, algae, and fungi
pyroclastic	pertaining to the fragmental ejecta released explosively from a volcanic vent
quartz	a common, igneous rock-forming mineral composed of silicon dioxide
radiolaria	microorganisms whose shells are made of silica and comprise a large component of siliceous sediments
radiometric dating	a process of age determination using chemical analysis of stable versus unstable radioactive elements

reef	biological community that lives at the edge of an island or continent in limestone deposits formed from the shells of dead organisms
regression	a fall in sea level, exposing continental shelves to erosion
reserves	known and identified energy and ore materials marked for imminent extraction and use
resources	reserves of useful Earth materials that might later be extracted
rhyolite	the volcanic equivalent of granite with abundant quartz and feldspar
ridge crest	the axis of a midocean spreading ridge aligned along the edges of two plates extending away from each other
rift valley	the center of an extensional spreading center where continental or oceanic plate separation occurs
roots, mountain	the deeper crustal layers under mountains
saltation	the movement of sand grains by wind or water
sandstone	a sedimentary rock consisting of cemented sand grains
saurian	a group of reptiles, including lizards, crocodiles, and dinosaurs
scarp	a steep slope formed by earth movements
seafloor spreading	a theory that the ocean floor is created by the separation of lithospheric plates along midocean ridges, with new oceanic crust formed from material that rises from the mantle to fill the rift
seamount	a submarine volcano
sedimentary rock	a rock composed of fragments cemented together
shield	areas of the exposed Precambrian nucleus of a continent
species	groups of organisms that share similar characteristics and are able to breed among themselves

spherules	small, spherical, glassy grains found in certain types of meteorites, lunar soils, and at large meteorite impact sites
strata	layered rock formations; also called beds
stromatolite	a calcareous structure built by successive layers of bacteria or algae; in existence for the past 3.5 billion years
subduction zone	a region where an oceanic plate dives below a continental plate into the mantle. Ocean trenches are the surface expression of a subduction zone.
submarine canyon	a deep gorge residing undersea and formed by the underwater extension of rivers
supernova	an enormous stellar explosion in which all but the inner core of a star is blown off into interstellar space
surge glacier	a continental glacier that heads toward the sea at a high rate of advance
symbiosis	the living together of two dissimilar organisms for mutual benefit
tectonic activity	the formation of the Earth's crust by large-scale movements throughout geologic time
tephra	all clastic (fragmentary) material, from dust particles to large chunks, expelled from volcanoes during eruptions
terrestrial	referring to all phenomena pertaining to the Earth
Tethys Sea	the hypothetical mid-latitude region of the world ocean separating the northern and southern continents of Laurasia and Gondwana several hundred million years ago
thecodonts	ancient, primitive reptiles that gave rise to dinosaurs, crocodiles, and birds
thermophilic	relating to primitive organisms that live in hot water environments
tide	a bulge in the ocean produced by the moon's gravitational force on the Earth's oceans. The rotation of the Earth beneath this bulge causes the rising and lowering of the sea level.

tillite	a sedimentary deposit composed of glacial till (non-stratified material deposited by glacial ice)
transgression	a rise in sea level that causes flooding of the shallow edges of continental margins
trench	a depression on the ocean floor caused by plate subduction
trilobite	an extinct marine arthropod, characterized by a body divided into three lobes, each bearing a pair of jointed appendages, and with a chitinous exoskeleton
tundra	permanently frozen ground at high latitudes
ultraviolet	the invisible light with a wavelength shorter than visible light and longer than X rays
uniformitarianism	a theory that the slow processes that shape the Earth's surface have acted essentially unchanged throughout geologic time
varves	thinly laminated lake bed sediments deposited by glacial meltwater
vascular	relating to channels for conveying bodily fluids
vertebrates	animals with an internal skeleton, including fish, amphibians, reptiles, and mammals
volcanism	any type of volcanic activity
volcano	a fissure or vent in the crust through which molten rock rises to the surface to form a mountain
wetland	land that is inundated by water and supports prolific wildlife
zooplankton	small, free-floating or poorly swimming marine or freshwater animal life

BIBLIOGRAPHY

PLANET EARTH

Black, David C. "Worlds around Other Stars." *Scientific American* 264 (January 1991): 76–82.

Boss, Alan P. "The Origin of the Moon." *Science* 231 (January 24, 1986): 341–345.

Hartley, Karen. "A New Window on Star Birth." *Astronomy* 17 (March 1989): 32–36.

Horgan, John. "In the Beginning." *Scientific American* 264 (February 1991): 117–125.

Kasting, James F., Owen B. Toon, and James B. Pollack. "How Climate Evolved on the Terrestrial Planets." *Scientific American* 258 (February 1988): 90–98.

Kerr, Richard A. "Making the Moon, Remaking the Earth." *Science* 243 (March 17, 1989): 1433–1435.

Newsom, Horton E., and Kenneth W. W. Sims. "Core Formation during Early Accretion of the Earth." *Science* 252 (May 17, 1991): 926–933.

Waldrop, Mitchell M. "Goodbye to the Warm Little Pond?" *Science* 250 (November 23, 1990): 1078–1080.

ARCHEAN ALGAE

Cairns-Smith, A. C. "The First Organisms." *Scientific American* 252 (June 1985): 90–100.

Horgon, John. "Off to an Early Start." *Scientific American* 269 (August 1993): 24.

Jones, Richard C., and Anthony N. Stranges. "Unraveling Origins, the Archean." *Earth Science* 42 (Winter 1989): 20–22.

Kerr, Richard A. "Plate Tectonics Is the Key to the Distant Past." *Science* 234 (November 7, 1986): 670–672.

Lowe, Donald R., et al. "Geological and Geochemical Record of 3400-Million-Year-Old Terrestrial Meteorite Impacts." *Science* 245 (September 1, 1989): 959–962.

Pool, Robert. "Pushing the Envelope of Life." *Science* 247 (January 12, 1990): 158–160.

Schopf, J. William, and Bonnie M. Packer. "Early Archean (3.3-Billion- to 3.5-Billion-Year-Old) Microfossils from Warrawoona Group Australia." *Science* 237 (July 3, 1987): 70–72.

Taylor, Stuart Ross. "Young Earth Like Venus?" *Nature* 350 (April 4, 1991): 376–377.

York, Derek. "The Earliest History of the Earth." *Scientific American* 268 (January 1993): 90–96.

Proterozoic Metazoans

Kerr, Richard A. "Another Movement in the Dance of the Plates." *Science* 244 (May 5, 1989): 529–530.

Knoll, Andrew H. "End of the Proterozoic Eon." *Scientific American* 265 (October 1991): 64–73.

Kunzig, Robert. "Birth of a Nation." *Discover* 11 (February 1990): 26–27.

———. "Horizontal History." *Discover* 10 (September 1989): 16–18.

McMenamin, Mark A. S. "The Emergence of Animals." *Scientific American* 256 (April 1987): 94–102.

Monastersky, Richard. "Oxygen Upheaval." *Science News* 142 (December 12, 1992): 412–413.

Weisburd, Stefi. "The Microbes That Loved the Sun." *Science News* 129 (February 15, 1986): 108–110.

Williams, George E. "The Solar Cycle in Precambrian Time." *Scientific American* 255 (August 1986): 88–96.

Cambrian Invertebrates

Beardsley, Tim. "Weird Wonders." *Scientific American* 266 (June 1992): 30–34.

Gould, Stephen J. "An Asteroid to Die For." *Discover* 10 (October 1989): 60–65.

Kerr, Richard A. "Evolution's Big Bang Gets Even More Explosive." *Science* 261 (September 3, 1993): 1274–1275.

Levinton, Jeffrey S. "The Big Bang of Animal Evolution." *Scientific American* 267 (November 1992): 84–91.

Monastersky, Richard. "Siberian Rocks Clock Biological Big Bang." *Science News* 144 (September 4, 1993): 148.

Morris, S. Conway. "Burgess Shale Faunas and the Cambrian Explosion." *Science* 246 (October 20, 1989): 339–345.

Palmer, Douglas. "Trilobite Fossil Shows Its Muscle." *New Scientist* 139 (November 6, 1993): 20.

Stolzenburg, William. "When Life Got Hard." *Science News* 138 (August 25, 1990): 120–123.

ORDOVICIAN VERTEBRATES

Bower, Bruce. "Fossils Flesh Out Early Vertebrates." *Science News* 133 (January 9, 1988): 21.

Forey, Peter, and Philippe Janvier. "Agnathans and the Origin of Jawed Vertebrates." *Nature* 361 (January 14, 1993): 129–133.

Howell, David G. "Terranes." *Scientific American* 253 (November 1985): 116–125.

Knox, C. "New View Surfaces of Ancient Atlantic." *Science News* 134 (October 8, 1988): 230.

Monastersky, Richard. "Trilobites: Not Forced Off the Block." *Science News* 142 (July 11, 1992): 30.

Svitil, Kathy A. "It's Alive, and It's a Graptolite." *Discover* 14 (July 1993): 18–19.

Weisburd, Stefi. "Facing Up to a Backwards Fossil." *Science News* 132 (July 18, 1987): 47.

SILURIAN PLANTS

Dickinson, William R. "Making Composite Continents." *Nature* 364 (July 22, 1993): 284–285.

Kerr, Richard A. "A Half-Billion-Year Head Start for Life on Land." *Science* 258 (November 13, 1992): 1082–1083.

Lewin, Roger. "On the Origin of Insect Wings." *Science* 230 (October 25, 1985): 428–429.

Monastersky, Richard. "Fossils Push Back Origin of Land Animals." *Science News* 138 (November 10, 1990): 292.

———. "Supersoil." *Science News* 136 (December 9, 1989): 376–377.

BIBLIOGRAPHY

Pestrong, Ray. "It's About Time." *Earth Science* 42 (Summer 1989): 14–15.

DEVONIAN FISHES

Barnes-Svarney, Patricia. "In Search of Ancient Shores." *Earth Science* 40 (Spring 1987): 22.

Eastman, Joseph T., and Arthur L. DeVries. "Antarctic Fishes." *Scientific American* 255 (November 1986): 106–114.

Kerr, Richard A. "Another Impact Extinction?" *Science* 256 (May 1992): 1280.

———. "How Long Does It Take to Build a Mountain?" *Science* 240 (June 24, 1988): 1735.

Landman, Neil H. "Luck of the Draw." *Natural History* 100 (December 1991): 68–71.

Richardson, Joyce R. "Brachiopods." *Scientific American* 255 (September 1986): 100–106.

Vogel, Shawna. "Face-to-Face with a Living Fossil." *Discover* 9 (March 1988): 56–57.

CARBONIFEROUS AMPHIBIANS

Fulkerson, William, Roddie R. Judkins, and Manoj K. Sanghvi. "Energy from Fossil Fuels." *Scientific American* 263 (September 1990): 129–135.

Hannibal, Joseph T. "Quarries Yield Rare Paleozoic Fossils." *Geotimes* 33 (July 1988): 10–13.

Monastersky, Richard. "Swamped by Climate Change?" *Science News* 138 (September 22, 1990): 184–186.

———. "Ancient Amphibians Found in Iowa." *Science News* 133 (June 25, 1988): 406.

Waters, Tom. "Greetings from Pangaea." *Discover* 13 (February 1992): 38–43.

Weisburd, Stefi. "Forests Made the World Frigid?" *Science News* 131 (January 3, 1987): 9.

PERMIAN REPTILES

Crowley, Thomas J., and Gerald R. North. "Abrupt Climate Change and Extinction Events in Earth History." *Science* 240 (May 20, 1988): 996–1001.

Kerr, Richard A. "Origins and Extinctions: Paleontology in Chicago." *Science* 257 (July 14, 1992): 486–487.

Murphy, J. Brendan, and R. Damian Nance. "Mountain Belts and the Supercontinent Cycle." *Scientific American* 266 (April 1992): 84–91.

Raup, David M. "Biological Extinction in Earth History." *Science* 231 (March 28, 1986): 1528–1533.

Shell, Ellen Ruppel. "Waves of Creation." *Discover* 14 (May 1993): 54–61.

Stanley, Steven M. "Mass Extinctions in the Ocean." *Scientific American* 250 (June 1984): 64–72.

TRIASSIC DINOSAURS

Benton, Michael J. "Late Triassic Extinctions and the Origin of the Dinosaurs." *Science* 260 (May 7, 1993): 769–770.

Buffetaut, Eric, and Rucha Ingavat. "The Mesozoic Vertebrates of Thailand." *Scientific American* 253 (August 1985): 80–87.

Burgin, Toni, et al. "The Fossils of Monte San Giorgio." *Scientific American* 260 (June 1989): 74–81.

Monastersky, Richard. "The Accidental Reign." *Science News* 143 (January 23, 1993): 60–62.

Morell, Virginia. "Announcing the Birth of a Heresy." *Discover* 8 (March 1987): 26–50.

Newton, Cathryn R. "Significance of 'Tethyan' Fossils in the American Cordillera." *Scientific American* 242 (October 21, 1988): 385–390.

Padian, Kevin. "Triassic-Jurassic Extinctions." *Science* 241 (September 9, 1988): 1358–1360.

Weisburd, Stefi. "Brushing Up on Dinosaurs." *Science News* 130 (October 4, 1986): 216–220.

JURASSIC BIRDS

Anderson, Alun. "Early Bird Threatens Archaeopteryx's Perch." *Science* 253 (July 5, 1991): 35.

Bonatti, Enrico. "The Rifting of Continents." *Scientific American* 256 (March 1987): 97–103.

Horner, John R. "The Nesting Behavior of Dinosaurs." *Scientific American* 250 (April 1984): 130–137.

Morell, Virginia. "Archaeopteryx: Early Bird Catches a Can of Worms." *Science* 259 (February 5, 1993): 764–765.

Nance, R. Damian, Thomas R. Worsley, and Judith B. Moody. "The Supercontinent Cycle." *Scientific American* 259 (July 1988): 72–79.

Robbins, Jim. "The Real Jurassic Park." *Discover* 12 (March 1991): 52–59.

Wellnhofer, Peter. "Archaeopteryx." *Scientific American* 262 (May 1990): 70–77.

CRETACEOUS CORALS

Alvarez, Walter, and Frank Asaro. "An Extraterrestrial Impact." *Scientific American* 263 (October 1990): 78–84.

Bird, Peter. "Formation of the Rocky Mountains, Western United States: A Continuum Computer Model." *Science* 239 (March 25, 1988): 1501–1507.

Courtillot, Vincent E. "A Volcanic Eruption." *Scientific American* 263 (October 1990): 85–92.

Hallam, Anthony. "End-Cretaceous Mass Extinction Event: Argument for Terrestrial Causation." *Science* 238 (November 27, 1987): 1237–1241.

Hildebrand, Alan R., and William V. Boynton. "Cretaceous Ground Zero." *Natural History* (June 1991): 47–52.

Monastersky, Richard. "Closing In on the Killer." *Science News* 141 (January 25, 1992): 56–58.

Morell, Virginia. "How Lethal Was the K-T Impact." *Science* 261 (September 17, 1993): 1518–1519.

Vickers-Rich, Patricia, and Thomas Hewitt Rich. "Polar Dinosaurs of Australia." *Scientific American* 269 (July 1993): 49–55.

TERTIARY MAMMALS

Coffin, Millard F., and Olav Eldholm. "Large Igneous Provinces." *Scientific American* 269 (October 1993): 42–49.

Harrison, T. Mark, et al. "Rising Tibet." *Science* 255 (March 27, 1992): 1663–1670.

Molnar, Peter. "The Structure of Mountain Ranges." *Scientific American* 255 (July 1986): 70–79.

Rampino, Michael R., and Richard B. Stothers. "Flood Basalt Volcanism During the Past 250 Million Years." *Science* 241 (August 5, 1988): 663–667.

Ruddiman, William F., and John E. Kutzbach. "Plateau Uplift and Climate Change." *Scientific American* 264 (March 1991): 66–74.

Storch, Gerhard. "The Mammals of Island Europe." *Scientific American* 266 (February 1992): 64–69.

Weisburd, Stefi. "Volcanoes and Extinctions: Round Two." *Science News* 131 (April 18, 1987): 248–250.

QUATERNARY GLACIATION

Anderson, P. M., et al. "Climate Change of the Last 18,000 Years: Observations and Model Simulations." *Science* 241 (August 26, 1988): 1043–1051.

Bower, Bruce. "Extinctions on Ice." *Science News* 132 (October 31, 1987): 284–285.

Broecker, Wallace S., and George H. Denton. "What Drives Glacial Cycles." *Scientific American* 262 (January 1990): 49–56.

Kunzig, Robert. "Ice Cycles." *Discover* 10 (May 1989): 74–79.

Matthews, Samuel W. "Ice on the World." *National Geographic* 171 (January 1987): 84–103.

Monastersky, Richard. "Ice Age Insights." *Science News* 134 (September 17, 1988): 184–186.

Stringer, Christopher B. "The Emergence of Modern Humans." *Scientific American* 263 (December 1990): 98–104.

Thorne, Alan G., and Milford H. Wolpoff. "The Multiregional Evolution of Humans." *Scientific American* 266 (April 1992): 76–83.

Wilson, Alan C., and Rebecca L. Cann. "The Recent African Genesis of Humans." *Scientific American* 266 (April 1992): 68–73.

INDEX

Boldface page numbers indicate extensive treatment of a topic. *Italic* page numbers indicate illustrations or captions. Page numbers followed by m indicate maps, by t indicate tables, and by g indicate glossary.